PHYSICS FOR ALL

PHYSICS FOR ALL

PHYSICS FOR ALL

Delve into the concepts of Kinematics, Thermodynamics, Relativity, Quantum Physics, and much more as a casual reader or as a physics student without the complexities of mathematics

ANISHA K. YEDDANAPUDI

Physics for All

First Edition

by Anisha K. Yeddanapudi

Copyright © 2020 Anisha K. Yeddanapudi

Published in 2020 by

Krishna Yeddanapudi

Email: kyeddanapudi@gmail.com

All rights reserved. No part of this book may be reproduced or transmitted in any form or by any means, electronic or mechanical, including photocopying, recording, or by any information storage and retrieval system, without permission in writing from the publisher.

CONTENTS

Foreword — viii
Preface — x
Introduction — 1
   Units and Measurement — 1
Kinematics — 5
   Mass — 5
   Inertia — 6
   Motion and Dimensions — 7
   Motion Measurement — 8
   Momentum — 10
   Angular Momentum — 11
Force and Motion — 13
   Newton's Laws — 13
   Dynamics — 16
   Newton's Laws Today — 29
   Gravitation — 29
   Kepler's Laws — 30
   Planetary Motion — 33
   Circular Motion — 34
   Satellites — 40
Work, Power, and Energy — 42
   Potential Energy — 42
   Kinetic Energy — 43
   Gravitational Potential Energy — 43
   Conservation of Mechanical Energy — 45
Fluids — 47
   Pressure — 47
   Density — 48
   Pascal's Law — 49
   Archimedes' Principle — 50
   Buoyant Force — 51
   Fluid Characteristics — 51

Bernoulli's Principle --- 52
Waves --- *54*
   Wave Parameters --- 56
   General Properties of Waves --- 58
   Elasticity --- 70
   Vibrations and Sound --- 75
   Gravitational Waves --- 88
Heat And Thermodynamics --- *90*
   Temperature --- 90
   Zeroth Law of Thermodynamics --- 91
   First Law of Thermodynamics --- 92
   Heat --- 93
   Entropy --- 101
   Second Law of Thermodynamics --- 102
   Third Law of Thermodynamics --- 104
   Kinetic Theory Of Gases --- 104
Electricity And Magnetism --- *109*
   Fields and Field Lines --- 109
   Electricity --- 112
   Magnetism --- 118
   Electromagnetism --- 121
   Electromagnetic Radiation --- 126
Light And Other Forms Of Energy --- *142*
   Light as a Wave --- 142
   Chemical Energy --- 146
   Nuclear Energy --- 147
Applications And Interesting Discoveries --- *149*
   Some Applications --- 149
   Interesting Discoveries --- 157
Theory of Relativity --- *159*
   Special Theory of Relativity --- 159
   General Theory of Relativity --- 162
Quantum Theory --- *166*
   Blackbody Radiation --- 166

Double-slit Experiments --- 167
Heisenberg's Uncertainty Principle --- 168
Södinger's Wave Equation --- 170
Quantum Entanglement --- 172
Quantum Computers --- 172
Photons --- 174

Matter and Particle Physics --- *176*
Atoms --- 176
Other Atomic Particles --- 182
Molecules --- 185
Dark Matter --- 187

Astronomy --- *189*
Solar System --- 189
Planets --- 191
Nearby Stars --- 201
Galaxies --- 204
Black Holes --- 209

Our Earth --- *212*
Atmosphere --- 212
Inner Layers --- 218

Interesting Theories of Physics --- *220*
String Theory --- 220
Problems with Assumptions --- 222
Space Radiation --- 227
Tesla Experiments --- 230
Interesting Conjectures --- 235

Bibliography --- *238*
Index --- *243*
About the Author --- *255*

FOREWORD

Throughout my life, I have always had a deep interest in physics and astronomy. Even as a child, my favorite toy was a planetarium projector. Yet, I never took a physics course until my Junior year of high school. Though I sincerely enjoyed the class and ended up learning a lot, I observed gaps within the curriculum that was being taught.

I also noticed that physics seemed to become a daunting subject among my peers. Numerous people either dropped the class or gave up during the semester because of fear and a lack of concepts. Nobody seemed to understand the beauty of the subject and everybody only focused on knowing the mathematical formulas.

To help remediate this issue, I decided to write this book. My goal was to provide every individual with the basic ability to get into the subject. Specifically, I wished for the concepts to be highlighted, and the mathematics kept to the side. Additionally, though we commonly learn about many widely known physics ideas and concepts such as Newton's laws, there isn't a focus on new concepts or more complicated ones.

Sadly, the only way one can seem to learn about quantum mechanics or string theory is through a thick, technical book filled with overcomplicated jargon and formulas that would make any casual reader confused and nervous even to try. So, I decided to fix this problem via this physics novel. Specifically, I chose to analyze overcomplicated topics like quantum mechanics or string theory without delving too deeply into their insanity. The goal was to keep everything simple and understandable for everyone.

Usually, the common practice in most high school physics classes is to skip the complicated concepts and focus on the formulas and problems. The issue with this approach is the large gaps of knowledge in understanding, which keeps lingering and building until one reaches a higher level college course. And by that time, it seems too late, and the dump of conceptual topics is incredibly overwhelming.

Additionally, during this time, I was also interested in Stephen Hawking's *A Brief History of Time*. This short novel managed to explain even the most complicated topics in the simplest of fashions, which made it easy for any reader, no matter their physics background, to pick up and learn. I thought about replicating a similar structure in my book and have been mostly successful.

I hoped that this book would also save some time for readers and learners. Usually, when trying to learn new concepts and ideas online, it takes time to find videos, watch them, and apply them. Additionally, some of these videos might not contain what you're looking for. And since we are all incredibly busy, reading a book like this saves time because everyone can quickly go to the dedicated chapter and find all of the concepts necessary. Thereby allowing us to circumvent the large videos and search process.

By reading this book, I hope that individuals can find a beauty in physics that would be difficult to spot. If even a single person purchases this book and re-evaluates or finds a new love for the subject, I would consider that a success. I sincerely wish that this book helps you find a passion you might not have had initially.

Anisha K. Yeddanapudi

PREFACE

The roots of Physics are to be found in the sixth century BC during the first period of Greek Philosophy. At that time, The Sages of the Milesian School in Ionia did not bother to separate science, philosophy, and religion. Their only quest was to feel and understand 'Physis,' which meant the nature of essential reality. To feel the essential reality, they practiced meditation, discourse, discussion, and debate. The root word of Physics then was Physis, which mainly consisted of meditation, and as a result, physics meant meditation.

From this organic frame of Physics, gradually, the subject changed into mechanics, the study of mechanical relationships based on technology, or techno-logic, the logic of inanimate machines.

The Classical or Newtonian Physics emerged due to the Philosophy of Rene Descartes, a 17th-century French Philosopher, who contended that all living forms are essentially lifeless machines, and only human beings are separate and distinct because God entered the human brain within the pineal gland. Although we know today that all life forms feel pain and emotion, the mechanical view of Rene Descartes, called the Cartesian view, which consisted of mechanical and mathematical logic, has become the basic paradigm of the science and scientific outlook is the Cartesian method. It consists of mathematical reductionism, or analyzing every phenomenon, into mechanically interacting parts amenable to mathematical equations. This view paved the way for the emergence of Classical or Newtonian Physics and the development of technology. The technological perspective abhors all emotions. Emotions lead to bias, and as a result, the study gets contaminated. The non-emotional, mechanical logic and mathematical reduction is even today the prized scientific method—the staunch opposite of spiritualism.

However, later after the development of Newtonian Physics, physics expanded its scope to include spiritualism. The results in quantum physics or the physics of the very small created many puzzles as the day-to-day experiences do not agree with quantum findings. Many of the findings of

Newtonian physics are yet to be proven in Quantum Physics, but the language of Newton is not suitable.

Many Quantum Physicists sound like Vedanthists, those who look to the Hindu Vedas for inspiration and enlightenment to get out of the quantum stalemates. The problem is quantum phenomena cannot be explained following the principles of Classical Physics. While the vast Microcosm (the universe of the very big, consisting of planets, stars, galaxies, quasars, etc.) is a gigantic puzzle, the area of Relativity is equally puzzling. The findings at the level of massive objects must be reconciled with the conclusions of the Quantum Universe.

The solutions are often found by certain physicists in Eastern Mysticism, to the chagrin of the scientists to whom spiritualism means upholding superstitions. Unfortunately these superstitions are what occur in reality.

As Physics today is wading in uncertainties, probabilities, assumptions, etc., In front of Physics, there are big mysteries and many unsolved questions. But the scope of Physics has become the Universe itself in all its dimensions. The greatest minds are engaged in creating agreement between the Quantum Physics and Relativity, The Strings Theory, The M theory family of Theories, the Loops Quantum Gravity Theory, and the Universe's view as a hologram, the study relating to the gigantic vacuum.

Physics has landed all of us in a permanent quest for answers, and one lifetime is not sufficient. Yet, in general, one would say that the journey might be better than the answers themselves. And I hope that there is a time shortly in the future where we will all look back and laugh at our overall cluelessness.

PHYSICS FOR ALL

INTRODUCTION

The original theories taught in every school physics textbook consist of several classical theories considered the building blocks of every high school physics class. Such views mainly include Newton's Laws, the Laws of Thermodynamics, and Kepler's Laws of Planetary Motion. Since their discovery at the beginning of the 18th century, these laws have governed most modern-day physics.

Later, in the 19th century, physicists realized that there is more to the Universe than those initially described by these laws. Today, with the constant changes in the modern era, such as quantum mechanics, the theory of relativity, string theory, quantum field theory, etc., one must realize that physics is no longer as simple as Newton determined in the early 18th century. Before delving into the craziness of neutrinos, photons, etc., let's take a step back in understanding the basics of a high school physics textbook.

After gradually progressing through the vast world of simple high school physics, we will be delving into the more complex topics, such as quantum mechanics, the theory of relativity, string theory, dark matter, Higgs Boson, elementary particles, etc. Finally, we will end with the newer views which have yet to be proven and are highly debated within the scientific community. The purpose of describing such contemporary theories is to allow everyone to understand the most complex topics and make one's own judgments. So, let's take a walk.

UNITS AND MEASUREMENT

Science and engineering are based on measurements and the comparison of these calculations. Hence, it becomes crucial to have rules about how things are measured. One aspect of physics is to create and conduct specific experiments that help establish the units for measurements and comparisons - thereby making this the basis of modern physics and incredibly essential to discuss.

MEASUREMENT

Each physical quantity is measured in its units with comparison to a standard. The standard corresponds to precisely 1 unit of a quantity. For example, the standard for a length corresponding to exactly 1.0 meter can be the distance that light travels in a specific length of time or based on some other measurement of three-dimensional space.

Hence, scientists must all agree upon a consistent standard, and the definitions of these standards must be sensible and practical. After setting up a standard, it becomes vital to work out systems by which any object can be represented through that standard. Rulers are one example of a procedure used to measure length; but, many comparisons need to be indirect. In other words, a ruler cannot be used to measure the radius of an atom or the distance from the Sun and back.

Due to the large number of physical quantities in the Universe, it becomes challenging to find unique units to define each of them. For this reason, we are lucky that many of these quantities are not independent and are based on one another. For example, speed is length over time. So, all we have to do is find a standard agreement for the quantities of length and time, and from there, define others based on these basic quantities. Almost all physical quantities can be derived through the multiplication or division of specific basic or fundamental quantities, known as derived quantities. For this reason, it becomes imperative to have base standards that are invariable and accessible for all.

FUNDAMENTAL UNITS

As mentioned earlier, most physical measurements are based on other independent quantities. These independent standards are fundamental for any other physically derived measurement. These entities are called fundamental quantities, and the units of these quantities are entirely separate from others. Such quantities include length, time, mass, temperature for heat, current in electricity, luminous intensity for light, and amount of matter or mole. The measuring units used for these fundamental quantities are called fundamental units. For instance, as we all know, we cannot express length in terms of time as both distance and time are entirely independent.

UNITS

A unit is a name that physicists assign to measures of a specific quantity. For example, the meter (m) is given to the amount of length. As stated previously, any unit that cannot be defined based on others is known as a fundamental unit. There are some main fundamental units such as length (meter), time (second), mass (kilogram), temperature (kelvin), current (ampere), luminous intensity (candela), and the amount of a substance (mole), to name a few.

All of these are defined based on specific standards to ensure consistency in experimentation. These quantities form the basis of the International System of Units, also known as SI units or the metric system. Various other systems take different fundamental quantities as base units, such as the CGS system or the FPS system. Yet, the SI system is the one most used in science and engineering worldwide [1].

SCALARS

Scalars are physical quantities which are entirely described by their magnitude. Examples of scalars include volume, distance, speed, mass, etc. Scalars are commonly characterized by real numbers, which are usually positive (occasionally, scalars can be negative, e.g., temperature, gauge pressure, etc.). Though scalars are a large aspect of physics, they are incredibly different from vectors.

VECTORS

Vectors are the basis of most physics and describe almost everything in physics, which has both a direction and magnitude. A vector's length represents the magnitude, while the arrow at the end represents its direction. In short, vectors have a direction component along with magnitude. This characteristic makes vectors integral for measuring advanced derived quantities like a magnetic field, electric field, etc.

Mathematically vectors are represented as arrows on plain paper. And using vector algebra, we can calculate the components of a vector in various directions. We can also do all the algebraic operations like addition, subtraction, multiplication, division, etc., using vectors, which can be important when analyzing motion in different directions.

Many properties of moving objects are described with vectors. Let's take a ball rolling on the ground; its velocity is represented by the vector's length, while the direction is marked with the arrow. The momentum of the ball is also defined as a vector quantity. Momentum vectors are useful to predict what happens when two objects collide with one another. And we will discuss this further in a future chapter. But the general idea is that vectors are a description of a quantity, using both magnitude and direction.

A brief note about this book, it only describes the concepts, and actual mathematical formulae, vector algebra, or calculus problems are not included. Sometimes simple algebraic mathematical formulas are portrayed in the diagrams to make the concept more clear. It is okay if you don't understand these mathematical formulas, as the written text does not require you to do so and only focuses on the concepts.

KINEMATICS

Kinematics is a part of classical mechanics and a branch of physics concerned with the motion of objects with the involvement of forces. Kinematics hopes to describe the position of material objects (displacement), the rate these objects are moving (velocity), and the rate at which their velocity is changing (acceleration). The use of kinematics to describe motion is only possible for particles with constrained actions (e.g., moving in pre-determined paths). A typical example of constrained motion would be a ball being thrown into the air. Everyone understands that the ball will reach a certain height and begin falling down, thereby creating a determined path for physicists to analyze. On the other hand, in space with free-motion, the forces determine the shape of the course or moving path.

An example of this would be throwing a ball into space. There is no determined path for the ball to move; it will continue floating in any direction where a force is applied. Scientists would commonly not use kinematics to describe such a motion. In this chapter, we will first be defining specific terms and then moving forward with the concepts themselves.

MASS

Many people use the terms mass and weight interchangeably. Note, in physics; there is a vast difference between the two of them. If you could count the number of protons, neutrons, and electrons in an object, this would measure mass. Simply put, mass is the amount of stuff that an item is made up of and is commonly defined in either grams or kilograms.

On the other hand, weight is a description of the gravitational interaction between objects that have mass. Newton described gravity as a force, so any object interacting with the Earth's gravitational pull, the strength of the interaction is known as weight. Hence, the units for weight are Newtons (the same as for forces). Additionally, weight and mass are proportional to each other. The weight of any object on Earth can be calculated by multiplying its mass with Earth's local gravitational field (whose value is 9.81 N/kg). But, this gravitational force is only applicable to objects on the surface of the Earth. Therefore, the gravitational force will be different on Moon or Mars.

Mass is an inherent property for all objects in our Universe. Yet, even today, scientists are still unsure about why this is a property. Researchers have tried to use data from particle colliders to explain the truth behind the mass. The most pronounced theory is the Higgs field and the particle Higgs Boson, which might give scientists a hint (we will discuss this in a later chapter), yet a lot more research needs to be done [2].

INERTIA

Like mass, inertia is another property of matter which opposes attempts to put an object into motion, or if it is moving, change the magnitude or direction of its velocity. Inertia enables a body to resist agents such as forces or torques. Yet, this idea commonly results in the common misconception that a body keeps moving because of inertia. This idea is false; the reason an object would keep moving is the absence of a force to slow it down, not due to inertia.

A clear example of inertia is when a train starts moving and a person standing within the train falls. The reason for the fall is because of the inertia of the individual. This person's body does not want to change from its restful state. But, when the train starts moving, there is a force being applied to the body. This force causes the individual to move backward since the train moves forward (Newton's third law). Hence, it is vital to hold the poles in a subway to ensure you do not fall down.

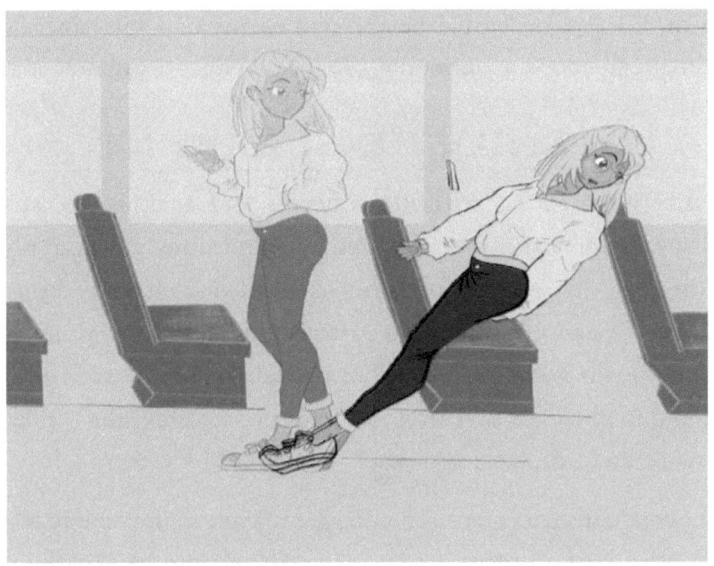

FIGURE 1: GIRL FALLING DUE TO INERTIA

MOTION AND DIMENSIONS

In physics, there are various ways an object can move: 1D (one-dimension), 2D (two-dimensions), and 3D (three-dimensions). One dimensional motion consists of objects moving in straight lines. This form of movement helps simplify complex physics problems into their essential components. An example of such motion would be a ball rolling on the floor, moving either right or left, in a straight line. Such a scenario would be a form of 1D motion.

On the other hand, a slightly more complex form of motion would be 2D. This form of motion consists of objects moving both in the y-direction (up and down) and the x-direction (right and left). A typical scenario would be throwing a ball up into the air at an angle; this would result in displacement in both the x and the y directions.

Finally, it is commonly known that we live in the 3D world, so there are scenarios where physicists must analyze 3D motion; this is where an object is moving in the x, y, and z directions. These types of problems get quite complicated, so usually, they are simplified to their 1D counterparts. An example of 3D motion would be the movement of a kite since it can be moving in all directions. Commonly, 3D motion occurs in outer space.

MOTION MEASUREMENT

Physics is based on scientific concepts and analysis; therefore, it is imperative to understand the measurements and units which apply. When first addressing motion, it becomes key to understand distance and displacement. Distance is somewhat common sense. Imagine running around a track. It is known that completing four circles around this track is equivalent to running a total length of 1 mile. Yet, after completing the 1 mile on this circular track, your displacement would be zero. Why is that?

Put simply; displacement is a change in position, not the total distance traveled. Also, note displacement is a vector quantity. Even though after completing four laps on the track leads to 1 mile, since you are right back where you started, the displacement is zero. There has been no change in position.

Like how the change in distance over time is known as speed, the change of displacement over time is known as velocity, which is also a vector quantity. The main difference between speed and velocity is that velocity is a vector value, so it has magnitude and direction. On the other hand, speed only has magnitude. While velocity tells you the speed and direction of an object, speed tells you how fast an item is traveling (without the direction).

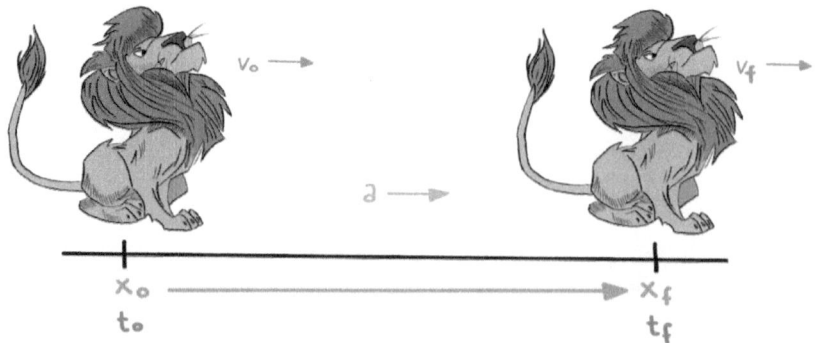

FIGURE 2: VECTOR AND SCALAR QUANTITIES FOR MOTION

We often hear the term accelerating in our day to day speech. Usually, it is assumed that acceleration means an increase in speed; this is a common misconception. Acceleration is a change in velocity over time, thereby also making it a vector quantity. Like velocity, acceleration also has both direction and magnitude. Since acceleration is a change of velocity over time, if an object is slowing down, there is negative acceleration, while if an object is speeding up, there is positive acceleration. When the object is reversing direction, velocity is zero at the instant the direction changes.

The terms displacement, velocity, and acceleration are used when an object moves in the x and y directions. The same terms also apply when something is moving in a circular path. Except there are a few fundamental changes.

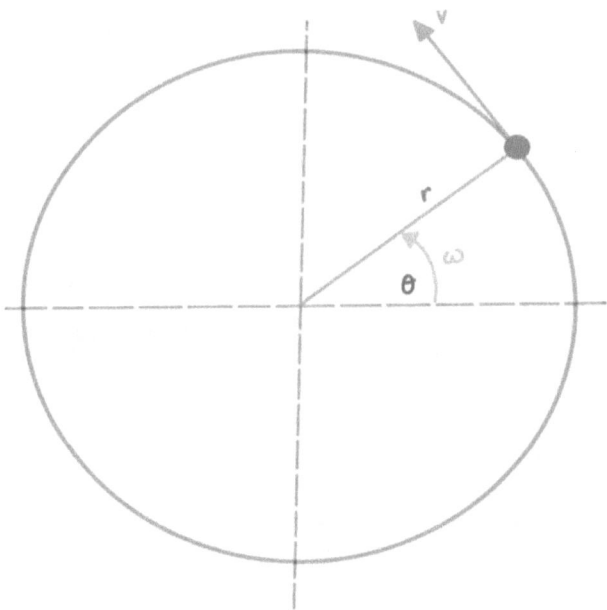

FIGURE 3: CIRCULAR MOTION MEASUREMENTS

When addressing displacement in the circular direction, we start using the angle θ (theta) to represent how much of the circle is covered within a rotation. Additionally, during circular rotation, velocity becomes known as tangential velocity. Also, to determine the angular speed an object travels in a circle, we start using ω (omega), and for angular acceleration, the symbol α (alpha). We will discuss these symbols' intricacies in a later chapter (circular motion), but for now, all you need to understand are the signs and their representation on a circle.

MOMENTUM

Momentum is another common term heard almost daily. For example, when the phrase "a lot of momentum" is used to describe sports teams or political candidates, one is commonly referring to the recent successes witnessed by the group or person. These recent successes usually mean that it will be challenging to change their trajectory in the future; this is the general essence of the word momentum when also used in physics. Quite plainly, momentum is a measure of a mass in motion or how much mass is in how much motion. Momentum is always a vector quantity and is described with the equation p = mass X velocity, where p represents the object's momentum.

FIGURE 4: MOMENTUM 'P' = MASS X VELOCITY

If an object has a lot of momentum, it is quite massive or moving at an incredibly high velocity and requires a large amount of force to change its speed or direction. Such as the lion depicted in the image above. The same can be said if an object has less momentum; it is easily moved with the smallest forces, such as a ball rolling across the floor.

Additionally, as shown with the momentum formula, a body's momentum is determined by both its velocity and mass. An object can have a massive momentum even if it has a smaller mass if it's traveling at high speed, such as an astronomical object the size of a grain of sand moving close to the speed of light in space. Even such a small object, due to its high speed, would be difficult to stop. Momentum is a crucial aspect of physics because it is conserved even when other measures in a system are not, which makes momentum useful when analyzing collisions and other such problems.

ANGULAR MOMENTUM

There are several objects in our Universe that move without moving, or, in other words, they move without going anywhere. Specifically, items that spin or rotate, such as a planet revolving around itself or an electron spinning. Still, just because they are not changing positions doesn't mean that they cannot be discussed. Like how momentum is how much "motion" an object has when moving in a straight line, angular momentum is how much "motion" an object has when going in circles or rotating.

Angular momentum is honestly quite simple. Think of a point, any point, and imagine an object moving in a circle around that one point. Then figure out how fast that object is moving around that circle. Multiplying its angular velocity and its moment of inertia would result in the object's angular momentum.

Angular momentum is an essential aspect of physics to address because when objects are interacting with each other, their angular momentums will not change over time. This idea can be shown through a simple thought experiment. The Moon has an enormous value of angular momentum when moving around the Earth in a circle (assume the orbit is a circle). If the Earth and the rest of the solar system were removed, the Moon's angular momentum would not change. Instead, the Moon would start spinning faster and move in a straight line away from the point that was originally the Earth.

This consistency of angular momentum is known as the Law of Conservation of Angular Momentum, which we will discuss later.

FIGURE 5: ANGULAR MOMENTUM 'L' = INERTIA X ANGULAR VELOCITY

FORCE AND MOTION

Newtonian physics, known as classical mechanics, is a branch of physics focusing on macroscopic objects' motion ranging from projectiles to parts of machinery. Classical mechanics is the foundation for physics and was the first branch of the subject discovered. This branch is mainly focused on Newton's applications and principles, specifically, his laws of motion [3].

NEWTON'S LAWS

As every high schooler can recite from memory, Sir Isaac Newton had three clear laws of motion that revolutionized how objects interact with one another. Before delving into the intricacies of these laws, it becomes essential to understand and analyze specific terms: external force and net force.

An external force originates externally or from the outside, rather than internally from the object. An example of an external force would be the Sun's gravitational pull on the Earth, while the force that the Earth's inner core exerts on the crust is an internal force since it originates internally. As you can see, it is just common sense. So, this moves us on to net force.

The net force on an object is the total force that is being exerted on the item. So, if several different forces act on the object, the net force would be the sum of all these forces. Since we have now cleared up some basic background knowledge, this moves us to the laws themselves.

NEWTON'S FIRST LAW (LAW OF INERTIA)

FIGURE 6: FREELY MOVING BODY

Every object remains in rest or constant motion unless acted upon by an outside force. There needs to be a cause, which is a net external force. Newton's first law can be thought of as an object's ability to maintain the motion status quo.

1. If one were to throw a stone, the force of gravity and the air resistance ensure that the stone stops moving over time. Suppose, if these forces were nonexistent, such as in space, the pebble would continue moving until it was acted upon by another force.

2. Additionally, a stone on the floor remains on the floor. As it is known, since the rock is at rest, it will not move unless a force is acted upon it.

NEWTON'S SECOND LAW

Force is equivalent to mass times acceleration. Newton's second law informs us that acceleration is proportional to the net force and inversely proportional to mass. So, if the net force is doubled, the acceleration becomes twice its original value. Also, if an object's mass is doubled, the acceleration is half its initial value if the external force remains unchanged.

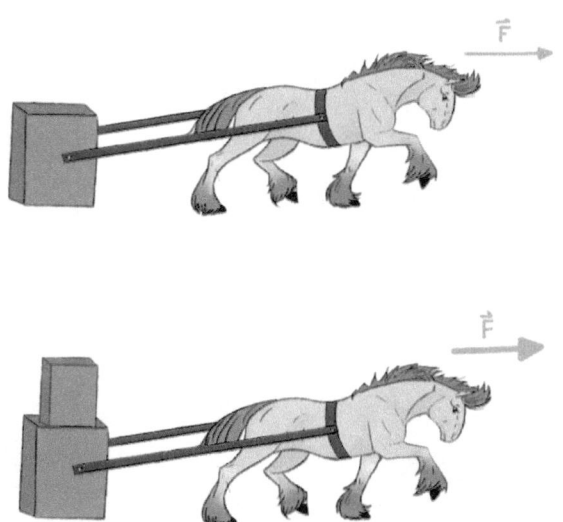

FIGURE 7: ACCELERATION REDUCED AS MASS IS HIGHER

1. If a satellite is orbiting in interstellar space at a constant velocity, no force needs to be applied for it to remain in motion. However, if the satellite wishes to change direction or increase its speed, it needs to add extra force (such as a thrust) to accelerate (either changing speed or direction).

2. When an object falls in free fall into the ground, you may notice that it increases its velocity as it's getting closer to the Earth. Or in other words, the object is accelerating. This is due to the force of gravity. Since gravity is the only force affecting an object during free-fall, it makes sense that acceleration is equivalent to acceleration due to gravity.

NEWTON'S THIRD LAW

For every action, there is an equal and opposite reaction. One body cannot experience a force without another body exerting a force. There is a form of action-reaction throughout nature. This law represents a level of symmetry within the universe: forces occur in pairs.

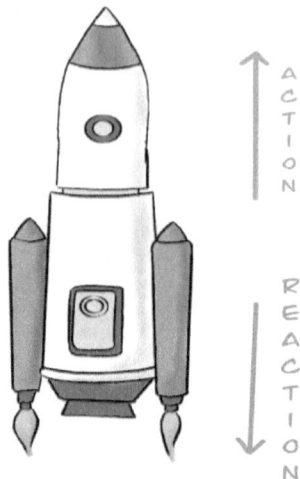

FIGURE 8: ROCKET MOTION PRINCIPLE

1. If one were to stand up, the reason the individual doesn't move up or down is that since the force of gravity pushes down, the normal force from the ground pushes up. The normal force is a reaction to the force of gravity, resulting in the individual remaining still.

2. Additionally, if a rocket were launched into the air, the force at the bottom resulting from the fuel thrusts the rocket up.

DYNAMICS

Dynamics is a branch of physics that is a subsection of mechanics concerned with the motion of particles and the physical factors that affect these particles ranging from force to energy. Galileo Galilei first laid out dynamics through experimentation with a smooth ball rolling down an inclined plane. Later, Isaac Newton elaborated on his ideas using his motion, momentum, and gravitation laws.

FORCE

Forces are, simply put, things that change an object's motion; they are commonly known as push and pull. Forces are also vectors; therefore, they have magnitude and direction. However, to clear up a misconception, forces are not properties of an object like mass or color. They are properties for an interaction between two objects; hence, it's impossible to have a force on an object without another object's involvement.

There are two different types of force: contact forces and action-at-a-distance forces. Contact forces are what they sound like; these forces require objects to be touching each other. Such forces include friction, air resistance, tension, etc. Action-at-a-Distance forces are forced that do not require contact for the force to act. Such forces include the gravitational force, the electrical force, the magnetic force, etc. So, for example, if you were to put a magnet near a metallic object, it doesn't need to touch the magnet for the object to be attracted to it.

FIGURE 9: EARTH'S FORCE OF GRAVITY

LAW OF CONSERVATION OF MOMENTUM

The law of conservation of momentum is what it sounds like; in an isolated system, the momentum of that system is conserved. It means the total momentum that one might start with would be equal to the system's final momentum. This law is commonly used to explain the velocity and mass of a collision along with other situations. Suppose two objects are on a collision course with one another. In that case, both objects' total momentum before the collision must equal the objects' total momentum after the collision. So, when two objects collide with one another, their velocity changes. This change in velocities will happen in such a way so that the momentum of the system is conserved. Additionally, there are some circumstances where energy might not be conserved; yet, momentum remains conserved in all collisions.

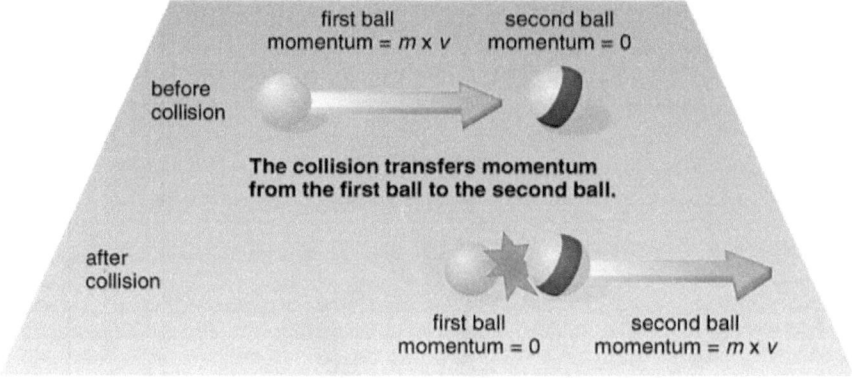

FIGURE 10: LAW OF CONSERVATION OF MOMENTUM

COLLISIONS

We have all probably witnessed two objects bumping into or colliding with each other. In physics, scientists commonly try to predict the outcome of scenarios, collisions being one of them. Predicting collisions can be quite useful for research into the physical phenomenon in both space and Earth.

There are two main categories of collisions: inelastic and elastic. There is no net loss of kinetic energy in the system due to the collision (both momentum and kinetic energy are conserved) in elastic collisions. Imagine two toy cars traveling towards each other with an equal velocity. They collide and bounce off each other, losing no net kinetic energy; this is an example of a perfectly elastic collision since no energy has been lost. Note, in our real world, perfectly elastic collisions are close to nonexistent. Though it is possible to have close to perfectly elastic collisions in specific circumstances, there is still a small amount of energy loss.

On the other hand, an inelastic collision occurs when there is a loss of kinetic energy; but, momentum is conserved. Now, this may seem unclear because we discussed that energy is always conserved. So, where did this kinetic energy go? It cannot disappear; that would be a violation of the First Law of Thermodynamics. The answer is simple. The kinetic energy is converted into something else, such as thermal energy, sound energy, or material deformation.

Suppose that you have two similar toy cars traveling towards each other. They collide; but, both vehicles are magnetized, and they stick together into one connected mass. This form of collisions is perfectly inelastic, which means the maximum amount of kinetic energy is lost. Yet, the final kinetic energy is not zero because momentum must be conserved.

Realistically, most collisions in the world are in between perfectly elastic and inelastic collisions. It is for this reason that modern vehicles are created to handle both inelastic and elastic collisions. The vehicle's frame is designed to absorb energy. Thereby ensuring the protection of the occupants inside the car.

ACCELERATION DUE TO GRAVITY

Acceleration due to gravity is quite simple: a body moving upwards stops gradually moving due to the influence of gravity, a body falls down with constant acceleration due to gravity and gravity alone. Note the acceleration due to gravity would be different on the moon in comparison to Earth. Since acceleration is a vector, it must have both magnitude and direction.

FIGURE 11: ACCELERATION DUE TO GRAVITY SAME ON ALL BODIES

Since the gravitational pull of the Earth or any planet is pointing towards the center of the planet, the gravitational force's direction is also downwards. Since the acceleration of a body is always taken in the order of the net force, and since the only force being considered in this circumstance is gravity, the acceleration should be downward (e.g., the direction of gravity).

When addressing the magnitude of g (acceleration due to gravity), the average value of g on the Earth's surface is 9.81 m/s². But, the Earth's gravitational pull increases as one moves closer to the Earth's sea level. So there would be a slightly larger amount of g at sea level than on top of Mt. Everest. Additionally, 9.81 m/s² is often confused with speed; but, this is not true. When an object is in free fall (the object is only affected by gravity), it accelerates at 9.81 m/s². By saying acceleration, we are saying that the speed is increasing at a rate of 9.81 m/s. So if a pebble is dropped, after one second, it's traveling at a speed of 9.81 m/s. After two seconds, the pebble's speed would be 19.62 m/s (which is equal to 2 X 9.81).

CENTER OF MASS

FIGURE 12: CENTER OF MASS

The center of mass is the average position of all parts of the system, weighted according to their masses. For simple rigid objects with uniform density, the center of the mass is located at the object's center. For example, the center of mass of a uniform disc would be at its center. On the other hand, there are times where the center of mass doesn't exist on the object. For a ring, the center of mass is at its center, where there isn't any material.

The intriguing thing about the center of mass is that it's a point where any force applied to the object acts. For example, the gravitational force of the Earth pulls from the center of mass of the object. Hence, if an object is subjected to an unbalanced force around its center of mass, it will rotate about its center of mass. The center of mass of an object is sometimes confused by the center of gravity, but they are not the same. The center of gravity is the point where the force of gravity acts on the object or system. At times this can be the center of mass; yet, depending on the object, it can be two separate points. Commonly, the center of mass and center of gravity are at the same point, and therefore the terms are used interchangeably, especially in classical mechanics.

PROJECTILES

Suppose you decide to throw a ball at an angle through the air. It takes a similar path to that of a parabola. In this case, the ball is considered a two-dimensional projectile since it's flying vertically and also horizontally through the air and only under the influence of gravity. Though all objects moving through the air experience some amount of air resistance, it is always assumed to be negligible to simplify the problem complexity. Though the amount of air resistance an object faces increases as its speed increases, we usually only address scenarios where the object's speed is small enough to consider air resistance negligible.

Since the gravitational force always pulls downward, gravity only affects the vertical component of the velocity vector. During projectile motion, the velocity vector's horizontal component remains constant as the ball moves through the air. When the ball reaches maximum height, the vertical component of the velocity is equal to zero. This is because the ball has momentarily stopped moving vertically at the maximum height and has to change directions to start falling down.

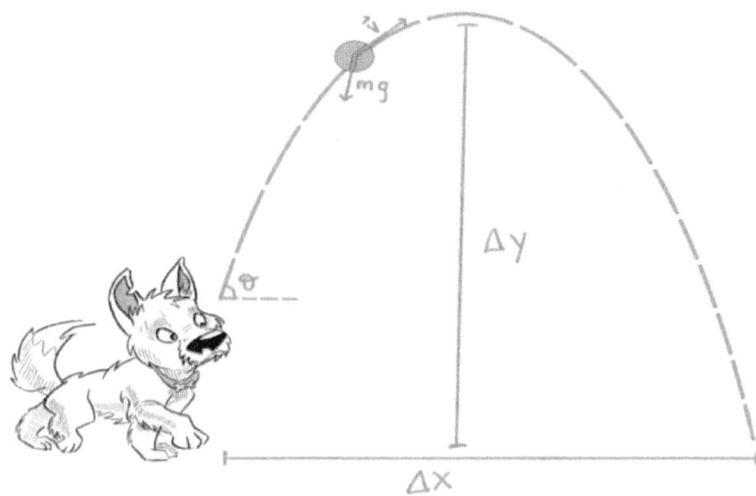

FIGURE 13: PROJECTILE MOTION IN TWO DIMENSIONS

When addressing projectile motion, it's imperative to take the two directions (x and y) separately. Therefore, one set of equations is used to describe motion in the x-direction, and another group is used to focus on the motion in the y-direction. The only reason we can do this separation is because the change in vertical velocity does not affect the horizontal velocity. So, though it may seem counter-intuitive, if you fire a bullet at an angle to the horizontal and drop another object from the maximum height reached by the bullet, both the objects will hit the ground simultaneously. This is because the horizontal velocity does not affect the bullet's vertical acceleration, so both the objects will change their vertical velocity by the same amount in the same amount of time.

There is no acceleration in the horizontal direction or x-direction during projectile motion since gravity doesn't pull on an object sideways. Though air resistance would, theoretically, cause a horizontal acceleration, slowing horizontal motion, since it's considered negligible in most cases, we assume a constant horizontal velocity for a projectile.

On the other hand, gravity does affect a projectile in the downward direction. So, there is a constant downward acceleration due to gravity for the projectile. As we know, since acceleration is constant; there is a particular set of kinematic formulas that can be used to solve for the specific values listed above. It's important to note when using these equations to only plug in the y component of the value on the y-direction and x-component of the value on the x-direction. Additionally, it is also important to note that initial velocity directed at an angle diagonally will have to be broken up into vertical and horizontal components to analyze each motion separately.

ROCKETS

Rockets are commonly used for space travel and are the basis of the majority of space research conducted by organizations such as NASA. One of the main aspects of a rocket would be its engine. A rocket engine burns fuel (either liquid or solid), turning it into hot gas, which is pushed out to provide the thrust necessary to lift the rocket. There is a difference between a rocket and a jet engine; while a jet engine uses air, a rocket engine doesn't require air (this is because a rocket engine has to continue to work in space, which is a vacuum.)

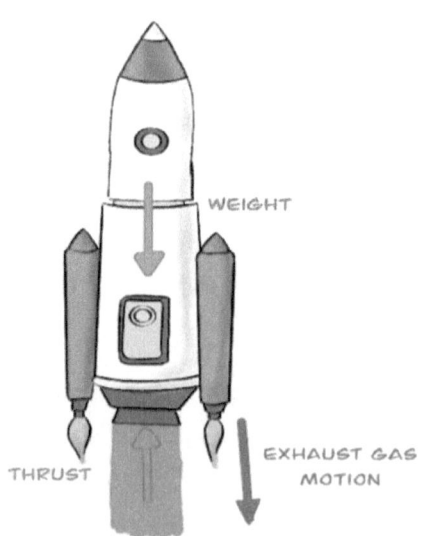

FIGURE 14: ROCKET MOTION

Once a rocket reaches space, there is nothing to push against, to move it forward. This causes the rocket to push exhaust, and because of Newton's third law of motion, the exhaust pushes back on the rocket. Therefore, allowing the rocket to continue moving forward in space as well.

EXAMPLES OF FORCES

Listed below are some forces that act on all bodies globally. These forces are not commonly visible, but we always experience their effect.

FRICTION

If you have ever gone on vacation to incredibly mountainous regions, parking your car to view the sites and be quite terrifying, especially on a steep incline. Still, we do so, anyway. We know that if we park our cars at a certain angle, they won't fall off the mountainside. Why is that? The answer is simple: friction. There are two major types of frictional force in physics: static friction and kinetic friction.

STATIC FRICTION

Static friction refers to the force between two surfaces, which prevents them from slipping and sliding across each other. This mainly applies when objects are static and unmoving and ensure that they remain this way when there is an opposite force pushing back.

For example, when you start to run, your planted foot can grip the ground and push back against it. This "grip" that your feet have to the ground is known as static friction and propels you forward. Therefore, if you have ever tried running on ice, you mainly end up jogging in place; this is because a smooth surface like ice has no static friction, so you cannot accelerate forward even if you were to plant your feet onto the surface.

The mathematical formula for static friction can be a bit difficult to understand because the friction force changes depending on the applied force on the object. Imagine trying to slide a heavy cardboard box across your living room. If you push it slightly, it will probably not move; you have to push with a strong enough force for it to budge and keep moving. So, it's easier to keep an object in motion than to start its motion (since inertia tells us objects like maintaining their state of rest). This means that an object's static friction will be larger than the kinetic friction because it takes more force to move an object at rest.

The equation for static friction is commonly written as the following inequality $F_s \leq \mu_s N$, since Fs' value can change and it can be as large as the quantity for $\mu_s N$. In this formula, F_s represents the force of static friction and N represents the normal force (which is the force that pushes the body against gravity, since as we learned in Newton's Third Law, forces come in pairs, in this case, the normal force is the reaction to a body's weight.

So, that leaves us with μ_s (coefficient of static friction). What is μ_s? To better understand μ_s imagine pushing a box across a rough surface like sandpaper. The box wouldn't start moving easily. Why? Simply put, the frictional force is higher, and it stops the box from moving. However, moving the same box across an ice surface would be easy for the same reason; there is a low static frictional force. This leads anyone to the logical conclusion that changing surfaces leads to different values of static frictional force. The symbol μ_s, also known as the coefficient of static friction, is a constant that considers the "roughness" of the two surfaces sliding across one another (the box and the ice in our example).

KINETIC FRICTION

If static friction describes the force of friction for a static object or to start the motion of an object at rest; then what is used to describe the frictional force of a moving object? To answer this, let's go back to the previous example of parking one's car on the edge of a mountainside. If someone were to accidentally push your car down the mountain, assuming that there were no boulders or rocks to stop its movement, it would come to a stop on its own over time. Why? The answer is described through the use of kinetic friction.

Similar to static friction, kinetic friction opposes the sliding motion of items and is responsible for reducing the speed of objects sliding across each other. Hence why, over time, your car, which is sliding down the mountain, will come to a stop. So, if there were no kinetic friction, your car would continue sliding forever and never stop.

The formula for kinetic friction is $F_k = \mu_k N$. The F_k represents the kinetic frictional force, and the N represents the normal force. This is almost the same as the formula for static friction. The only difference is that equal sign instead of an inequality. The reason being that kinetic friction does not change its quantity depending on the force applied. It remains constant to a large degree because the object has already been put into motion, which is not the case for static friction. The variable μ_k is the coefficient of kinetic friction and is generally less than the value of the coefficient of static friction μ_s.

AIR RESISTANCE ON EARTH

Air resistance (drag) is a form of friction that occurs when a body passes through the air. An excellent way to look at air resistance is due to an object's surface colliding with air molecules. This means that the two aspects that have the most considerable impact on air resistance would be the object's velocity and cross-sectional area. All these factors are directly proportional, so increasing either cross-sectional area or velocity would directly increase air resistance.

Overall, there are three main categories of air resistance: lift induced, parasitic, and wave. Lift induced is a consequence of drag being created on the body due to fuel or wings. On the other hand, parasitic drag is a form of air resistance that occurs when a body moves through a fluid (gas or liquid), similar to a ball falling from up in the air. As the object slows down in the fluid, the effect of parasitic drag increases, leading to specific airspeeds where this drag has minimal if not no effects. Finally, wave drag is quite similar to parasitic drag and results from a body moving incredibly fast through a compressible fluid. Commonly, wave drag occurs at speeds around Mach 0.5 (half the speed of sound); still, these speeds have to be less than Mach 1. The reason for wave drags in these circumstances would be supersonic flow, which are shockwaves forming at the tip and tail ends of the body moving at very high speeds.

Understanding air resistance is a critical aspect of aerospace travel and allows us to develop technology with the ability to move throughout Earth's atmosphere and also space.

GRAVITY AND WEIGHT

As stated in the previous chapter, weight is simply another way to represent the planet's gravitational pull. Since the Earth pulls all objects on its surface towards its core, weight is the force of the Earth's interaction and pull. You calculate weight by multiplying the object's mass with the acceleration due to the Earth's gravity, which is 9.81 m/s^2.

FIGURE 15: ALL BODIES EXPERIENCE EARTH'S GRAVITY

All objects are affected by Earth's gravitational pull, no matter their position. This force ensures that you do not keep floating away into outer space. It allows you to remain and stay on Earth. Yet, the force of gravity decreases the further one moves from the Earth itself. Even the international space station where individuals experience weightlessness still faces a gravitational pull of 8.7 m/s^2. Hence, objects always are in free-fall orbit around the Earth.

NEWTON'S LAWS TODAY

Before creating Newton's laws, objects in motion were mostly unexplained phenomena. Through his laws, Newton tried to explain every object's movement in the universe, and to a large extent, he was successful.

Note, in the early 20th century, with the discovery of atomic and quantum particles, these laws no longer applied. This was a massive shock to the scientific community, showing that nothing in our universe is ever certain. These particles were moving differently than what could be predicted by Newton's laws, therefore new laws and theories needed to be created on the quantum level.

Newton's laws can be used to derive a series of other laws, such as the law of conservation of momentum for linear motion and the law of conservation of angular momentum for circular motion. In general, Newton's laws can be used to explain and calculate parameters of movement such as speed, acceleration, force, etc., and all types of motion for large objects.

GRAVITATION

All bodies in the universe (stars, planets, galaxies, etc.) are attracted to one another; this force is known as the gravitational force (or simply gravity). This force was, initially, discovered by Newton along with a formula on how to calculate it. According to this formula, the gravitational force is proportional to the bodies' mass and inversely proportional to the square of the distance between these bodies (formula for gravitational attraction between two masses M_1 and M_2: $F_G = G.M_1.M_2/R^2$, where G is a gravitational constant).

The force of gravity is quite commonly noticed by one's weight (or the attractive force that an individual has to the Earth). The actual reason for the existence of forces in the universe is unknown, including gravity. Observation has proven that these forces exist; but, their "origins" in our Universe are still not completely understood. As can be expected, there are theories for the existence of gravity, one of the most famous ones being that of Albert Einstein, who described gravity's existence due to the curvature or bending of space-time near heavy objects (this will be discussed further later in this book).

KEPLER'S LAWS

Kepler's laws of planetary motion are consistently used to determine the movement of large bodies in the solar system. To better understand Kepler's laws, two key geometric shapes come into play, the circle and the ellipse. An ellipse has two focus points called foci. All planets, asteroids, moons, etc., move in elliptical orbits. An oval is inherently an ellipse. A circle has a single-center, thereby making it a particular case of an ellipse where both centers are at the same location [4].

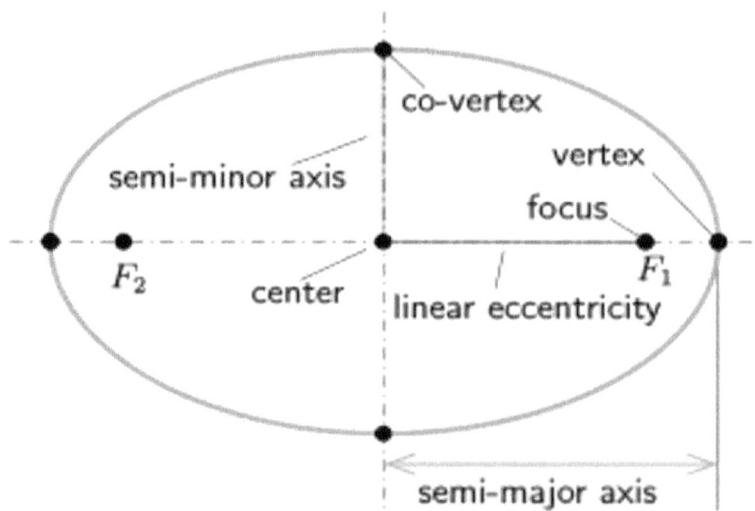

FIGURE 16: ELLIPSE WITH FOCI F1 AND F2

KEPLER'S FIRST LAW (THE LAW OF ORBITS)

All planets move in elliptical orbits, with the Sun at one of the foci. Hence this makes it easy for us to determine the path of the movement of planets at any particular instance of time. This law is used extensively in calculating the orbits of objects in our solar system.

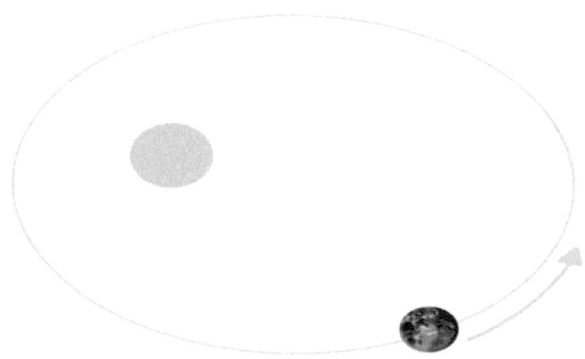

FIGURE 17: PLANET IN ELLIPTIC ORBIT AROUND THE SUN

1. This law's phenomenon is quite commonly seen in the night sky and several planetary diagrams in an extensive series of physics textbooks. Consistently, upon observation by several telescopes and satellites, the Sun is the center of the solar system, and the planets move in elliptical orbits around it.

2. Imagine a ball with a string tied to it. And if one were to take the rope and rotate it, the ball would move in a circle with the individual's hand being at the center. Now, if this hand were the Sun, the ball was a planet, and the string is the gravitational force between them. This is a clear example of how our solar system works under Kepler's first law.

3. There are only two ways that planets can move in the solar system, either in a circle (which usually does not happen) or in an ellipse. If no other objects were attracting the planet Earth in our solar system, the Earth would have a circular orbit. Since several other objects are pulling at the Earth throughout our solar system, the Earth has an elliptical orbit. This elliptical orbit is also true in the case of other planets, such as Jupiter or Saturn.

KEPLER'S SECOND LAW (THE LAW OF AREAS)

A line that connects a planet to the Sun sweeps out equal areas in equal times. This law helps us in calculating the speed of movement for planets at any instance of time.

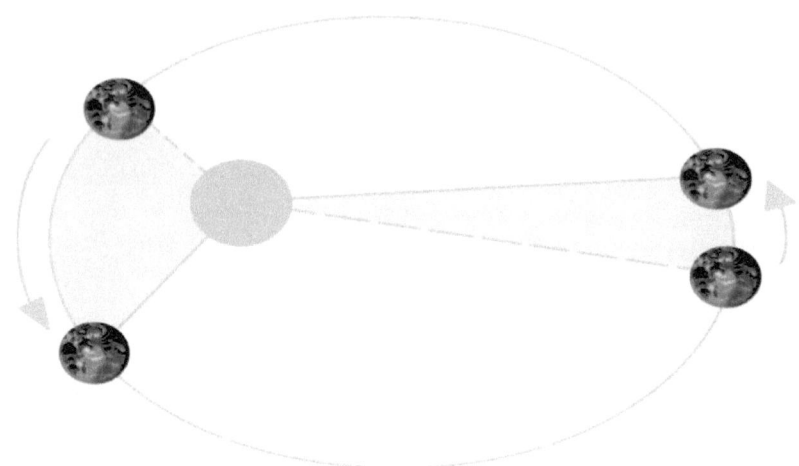

FIGURE 18: AREAS SWEPT BY THE PLANET

1. Simply put, this law states that when planets are closer to the Sun, they move faster than when they are farther away. Additionally, through this law, one can calculate a planets' speed at different points in their orbit.

2. By using the observations of Copernicus, Kepler was able to formulate these laws. The exact reason why planets move in this specific orientation is still not completely understood; yet, this form of motion is observed in all planets.

KEPLER'S THIRD LAW (THE LAW OF PERIODS)

The square of the time period (the time taken to complete one revolution about the Sun) of any planet is proportional to the cube of the semi-major axis of its orbit.

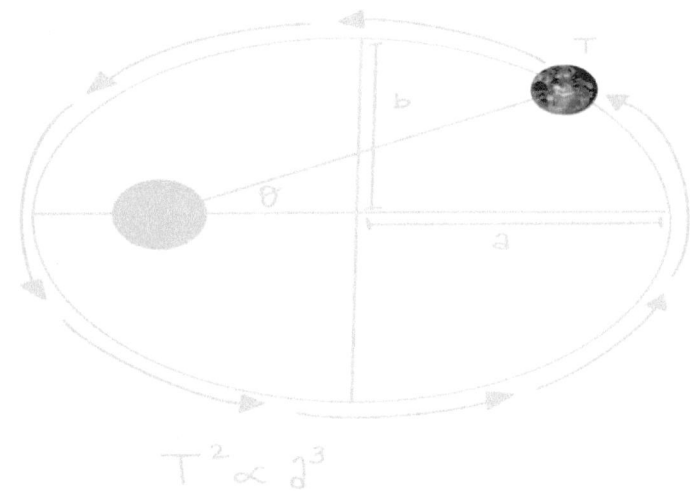

FIGURE 19: SQUARE OF THE TIME PERIOD PROPORTIONAL TO THE CUBE OF SEMI MAJOR AXIS

1. Quite simply, this law provides a way to calculate the period of rotation of planets in orbit, along with the size of the orbit's ellipse.

2. The reason for this kind of motion and the existence of gravity among planets is still unknown, as stated above.

PLANETARY MOTION

FIGURE 20: SOLAR SYSTEM WITH SUN AND PLANETS [5]

Planets in our solar system orbit the Sun in a counterclockwise direction. These orbits are aligned with what is known as the ecliptic plane. An ellipse is based on two points, each called foci. The sum of distances from the foci to any point on the ellipse is constant. Additionally, the "flatness" of an ellipse is known as eccentricity (the eccentricity of an ellipse ranges from 0, a circle, to 1). The final property is that the longest axis is called the major axis, while the short axis is called the minor axis.

CIRCULAR MOTION

Motion cannot always be in a straight line; that would be boring. Imagine going on a rollercoaster that only travels straight. The experience would be lackluster at best. It is for this reason we also have circular motion. Hence, we need to define a different set of measuring quantities to describe circular motion since it differs slightly from normal straight motion.

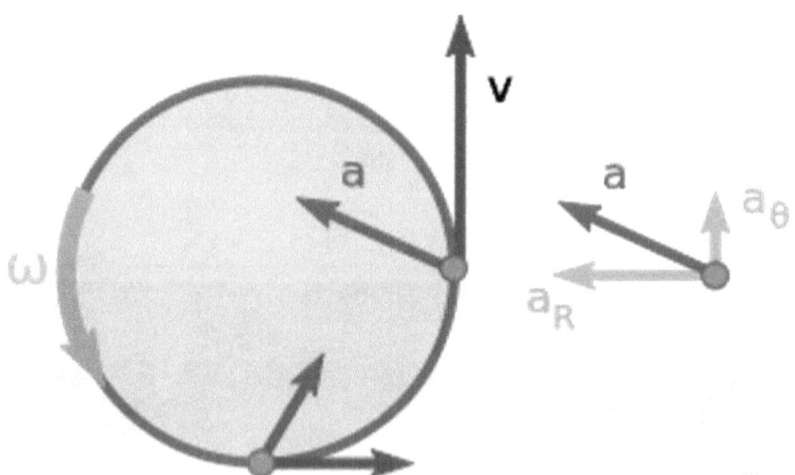

FIGURE 21: CIRCULAR MOTION PARAMETERS [6]

Even if an object is traveling in circles, it is still covering distance; this is described through the angle traversed (every time one rotation is completed, the angle traversed is 360 degrees or 2pi radians). Displacement is defined as the angular displacement, which is also a vector quantity. The movement of the body determines the direction of angular displacement (clockwise or counterclockwise).

Angular velocity (ω) measures the number of rotations that take place over time. The vector corresponds with either a clockwise or counterclockwise direction. Angular speed is the absolute value of angular velocity, so the same symbol (ω) without the vector bar is commonly used to represent both. On the other hand, linear velocity (v) is represented with the amount of displacement over time, where linear speed is the positive value for linear velocity. Like linear motion, we can derive a whole set of formulae for circular motion. These formulas seem incredibly like those of linear motion, and the parallels between them are almost uncanny.

LAW OF CONSERVATION OF ANGULAR MOMENTUM

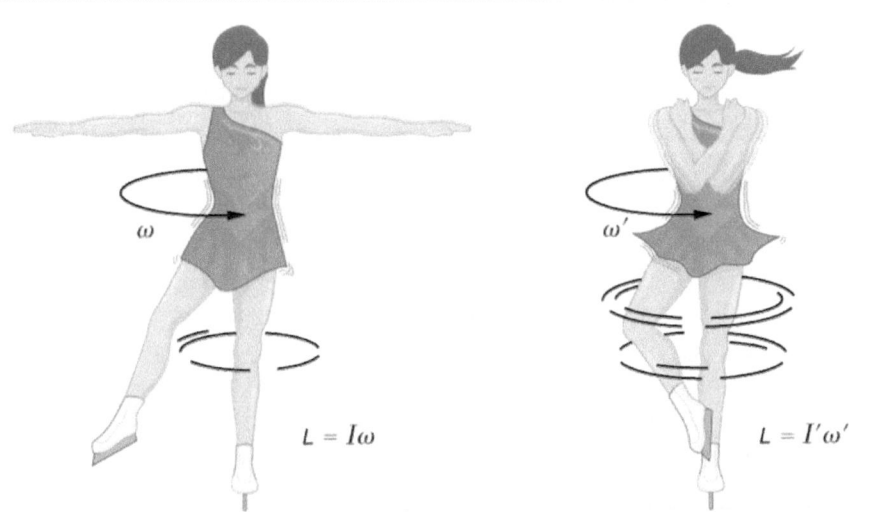

FIGURE 22: CONSERVATION OF ANGULAR MOMENTUM [7]

Similar to how linear momentum is conserved, the same law is also applied to angular momentum. There are a lot of parallels between linear motion and circular motion, as you might have noticed.

Angular momentum depends on both an object's rotational velocity along with its rotational inertia. When an object changes its shape or the value of its rotational inertia, its angular velocity also changes (if there is no external torque). For example, when an ice skater spins and increases her rotational speed by pulling her arms inward. The reduced rotational inertia (due to the smaller shape) has to be compensated through the increased speed to conserve angular momentum.

In an example given previously, a system of planets orbiting a star maintains a constant angular momentum since there is no net external torque. This is why a planet's speed is reduced when it's further away from the star; however, it increases when it approaches the star—thereby maintaining the angular momentum around the star by consistently adjusting the velocity and distance of the planet and the star.

TORQUE

Anyone who has ever opened a door before has a simple understanding of torque. When a person opens a door, they push on the door's side, which is farthest from the hinges. If one were to push closer to the hinges, it would require more force, even though the same amount of work is done for both scenarios. The force that causes the door to turn is known as torque.

Torque measures the force that causes an object to turn about an axis. Similar to how a force causes an object to accelerate in linear motion and kinematics, torque causes an object to accelerate in the angular direction. Like all forces, torque is a vector, and the direction depends on the force on the axis.

Torque can be either static or dynamic. Static torque does not produce angular acceleration. Someone pushing on a closed door is applying static torque to the door since it is not rotating, despite the force being applied. When a force causes the object to accelerate, this is known as dynamic torque. So turning and pushing the door handle would result in dynamic torque because the door is accelerating around its hinges.

The following equation is used to calculate torque:

$T = F.r.sin(A)$ *(Torque = Force X Perpendicular Component of Radius)*

"R" is the length of the moment arm, the moment arm is the radius at which the force acts on the object, and 'A' is the angle between the force vector and the moment arm. Additionally, through this equation, you will notice that the longer the radius (the moment arm), the larger the torque generated. Hence, why door handles are placed farther away from the hinges to produce the most torque necessary.

Since torque is a vector quantity, it also has direction, both positive and negative. If the torque causes the object to move in the counterclockwise direction, the force is positive, while it is negative if the object moves in the clockwise direction. Additionally, Newton's second law can also be applied to rotational kinematics in the form of torque.

$T = I.a$ *(Torque = Inertia X Angular Acceleration)*

Notice the similarities between this and F = ma. Here 'a' is the angular acceleration, and "I" is the rotational inertia, which is the property of a system that depends on the distribution of mass within it.

FIGURE 23: TORQUE FOR REMOVING THE CAR TIRE

When addressing torque, it's also essential to address the idea of rotational equilibrium. The concept of rotational equilibrium has the same principle as Newton's first law. If an object is not rotating, it remains not rotating unless acted on by an external torque. Similarly, an object rotating (or spinning) at a constant angular velocity remains rotating (or spinning) unless acted on by an external torque. This concept is essential when addressing multiple torques in a system. Usually, in this circumstance, we address the net torque to determine the system's different values.

Just like how all forces relate to power and energy, torque is no different. Torque and energy have the same fundamental units; but, they do not measure the same thing. Torque is a vector quantity that can only be defined as a rotatable system. Power can be calculated from torque if the rotational velocity of the system is known. The horsepower of an engine is calculated similarly by measuring torque and rotational speed.

CENTRIFUGAL FORCE

Centrifugal force is commonly confused with its counterpart, centripetal force. Yet, they are two sides of the same coin. Centripetal force is necessary to have an object continue moving in a curved path and is directed inward towards the circle's center. Since forces come in pairs, centrifugal force is the apparent force felt by the object but acts outwardly from the circular path of rotation. This makes centrifugal force and centripetal force similar to the normal force and the gravitational force. One pushes outward while the other inward.

Still, the centrifugal force is based on one's reference point. So, if we were to imagine a mass attached to a string, the string would extent an inward centripetal force on the mass, and the mass appears to exert an outward centrifugal force on the rope. Centripetal and centrifugal are the same force; they are just viewed from different reference points. If you are considering a rotating system from the outside, you will view a centripetal force. On the other hand, if you are experiencing a rotation, such as going in a loop on a rollercoaster, you would experience a centrifugal force pushing you out of the coaster., even though it's a centripetal force ensuring that you don't fall out.

EQUILIBRIUM

Equilibrium is the condition when neither a system's state of motion or its internal energy state tends to change with time. A mechanical body is considered to be in equilibrium if it experiences neither linear nor angular acceleration. This object will continue in this condition indefinitely unless acted upon by an outside force (as stated by Newton's first law).

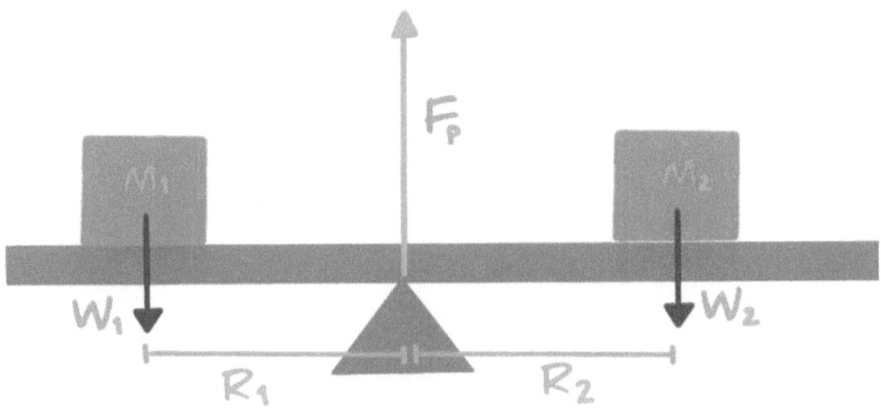

FIGURE 24: EQUILIBRIUM AS 'W1 X R1 = W2 X R2'

Additionally, the vector sum should be equivalent to zero for a rigid body to be in equilibrium. A common misconception is that equilibrium means that the body is not moving. Though this can be true, a body can also be in equilibrium if traveling at a constant velocity (not accelerating).

In thermodynamics, equilibrium is extended to induce the possible changes in a system's internal state ranging from temperature to density. At thermodynamic equilibrium, the system's temperature is uniform, and any other external forces are balanced so that they remain constant. At times, the heat flow in a system can also remain stable under these circumstances.

SATELLITES

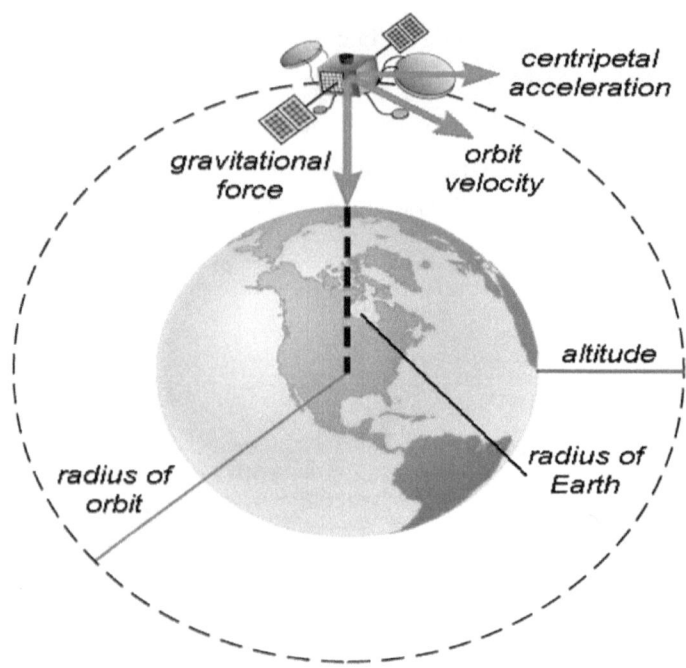

FIGURE 25: SATELLITE MOTION AROUND THE EARTH

Satellites are always in constant orbit around the Earth. There are probably hundreds of satellites surveying and photographing the Earth and planetary bodies hundreds of thousands of miles away. The way that satellites orbit varies widely; however, a couple of significant laws help govern their function in the exosphere.

GEOSTATIONARY ORBIT (GEO)

A satellite in geostationary orbit is in a fixed position when viewed by an Earth-based observer. This satellite revolves around the Earth at a consistent speed and is commonly used for communication. Ground-based antennas are directed toward the satellite and can operate without equipment to track the satellite's motion. These satellites commonly orbit above the equator but are not always suitable for providing communication at high altitudes.

MEDIUM EARTH ORBIT (MEO)

The medium Earth orbit or MEO is the area in between the two most popular satellite orbital zones. Only around 10% of satellites orbit in this area. The reason for this are the Van Allen Radiation Belts. These belts are two doughnut-shaped rings where Earth's magnetic field traps cosmically charged particles. These particles result in these areas being highly radioactive and can damage satellites frequently. The majority of satellites in this area are fitted with shielding to minimize the impacts; yet, the problem persists.

POLAR ORBIT

Specific LEO (Low Earth Orbit) satellites that orbit relatively close to the Earth's surface also follow the polar orbital path; this means that they pass over the poles while completing one full orbit around the Earth instead of traveling horizontally via the equator. Due to Earth's rotation, these satellites do not have a clear pre-determined path and pass over different swaths of Earth's surface with each revolution they take. This type of orbit's main benefit is a satellite's ability to inspect every point in its orbit twice in one 24-hour day; this makes such an orbit integral for analysis-based satellites who check the weather, storms, crop yield, etc.

WORK, POWER, AND ENERGY

Energy is a word that we commonly tend to use a lot in everyday life. Even though it's used quite loosely, there is a scientific definition for it. Energy is a measurement of the ability to do work. This means energy itself is not a material substance; yet, it can be stored and measured in many forms.

Though there is the concept of energy consumption, energy is never created or destroyed. Energy is transferred from one form to another and does work in the process.

There are many different types of energy: the two main ones are kinetic energy and potential energy. Kinetic energy is the energy associated with an object's movement, and potential energy is the energy that an item has based on its position and the other objects around it.

Suppose you ate a candy bar and wanted to burn off the calories. The way that you would go about burning these calories would be through work. Work is considered to be the transfer of energy. So when you want to burn calories, you would take that energy, do work (exercise, walk, run, etc.) and transfer it into some other form of energy, whether it's kinetic or thermal—thereby losing the calories and fat gained from the candy bar.

POTENTIAL ENERGY

As stated previously, there are two primary forms of energy, potential or latent energy when an object is at rest and kinetic energy when an object is in motion. Potential energy (PE) is quite simply what it sounds like; it is the energy that an object potentially has. The actual potential energy of an object depends on its position relative to other things in a system. We will discuss gravitational potential energy later in this chapter. Since the measurement of any given object's potential energy is a measure of its potential to do work, you would calculate the total potential for work for all things within the system. This concept broadly refers to gravitational potential energy, though there are other forms of potential energy.

KINETIC ENERGY

Kinetic energy (KE) is the energy that is an object that has because of its motion. If one were to accelerate a mass, then a force needs to be applied. Applying a force requires us to do work; therefore, energy needs to be transferred to the mass to make the object move at a constant speed. The transferred energy is known as kinetic energy, and that depends on the mass and speed achieved.

Kinetic energy and velocity have a quadratic relationship, which means that as velocity doubles, kinetic energy quadruples. Additionally, kinetic energy must always equal zero or positive; there cannot be negative kinetic energy. Also, energy is not a vector quantity; it only has magnitude; there is no direction associated with energy.

Kinetic energy is also related to work because the change in kinetic energy is equivalent to the system's (or object's) net work done. This idea is expressed as the work-energy theorem and generally applies with forces that vary in both direction and magnitude. It is crucial to notice this to determine the conservation of energy in a system.

GRAVITATIONAL POTENTIAL ENERGY

All forces can create potential energy (since all forces can do work). Since gravity is also a force, it's no exception. Gravitational potential energy represents the potential an object has to do work due to being located at a particular position in the gravitational field.

FIGURE 26: BALL'S POTENTIAL ENERGY CONVERTING TO KINETIC ENERGY

Imagine a ball lifted to a height (h) against the force of gravity (upwards), as shown in the image above. Using the work equation, we can calculate the gravitational potential energy by multiplying the height with the gravitational force. This formula assumes that the acceleration due to gravity is constant over the height.

If the force holding the ball up were removed, the object would fall back down to the ground. This is an example of gravitational potential energy being converted into kinetic energy (as the ball is falling downwards). Since energy is always conserved, the total gravitational potential energy will always be equal to the total kinetic energy (mathematically: PE = KE or mgh = $mv^2/2$, where v is the final velocity before touching the ground).

The intriguing thing about gravitational potential energy is that any vertical location can be taken where the height is equivalent to zero. Therefore sea level doesn't necessarily equal zero. Depending on the circumstance, the top of Mt. Everest can be taken as the base location of h=0. This reference point for a given problem is sometimes referred to as a datum. Hence, the gravitational potential energy could be negative if that object were to pass below the reference point. As long as that point is consistently taken as constant, there should be no issue among calculations.

CONSERVATION OF MECHANICAL ENERGY

Conservation refers to something remaining constant and unchanging. In the physics sense, this means that before and after an event has occurred, this specific quantity remains unchanged. There are various conserved quantities in physics, which become incredibly useful in predicting scenarios. In this section, we will be discussing the overall conservation of mechanical energy in a system. Hence it means that the energy might change its form; but, the total amount of mechanical energy will be conserved as there are no other types of energy interactions like heat, light, etc.

Additionally, the conservation of mechanical energy only applies to isolated systems. Suppose, you have a ball rolling across the floor. In this circumstance, the conservation of energy doesn't apply because it's an open system. Unless both the floor and the ball are considered in the system, then the law of conservation of energy applies.

For clarification, a system in physics refers to a collection of objects in which the problem takes place. The things outside of the system are not considered in the problem and are collectively labeled as the environment. Though ignoring the environment can result in a slight amount of inaccuracy in the equations, it ensures that there is an amount of simplification and is a common practice within physics. When defining a system, physicists focus on the main object within the problem and all other entities that this main object would interact with.

Now, mechanical energy is the sum of the potential and kinetic energy within a system. Note, potential energy is only applicable to forces like gravity or the spring force, which are conservative. It is inapplicable for nonconservative forces such as friction or air resistance because the energy of these forces usually dissipates into the environment in the form of heat. This means that the conservation of mechanical energy mainly applies to conservative forces; yet, it is still quite useful in solving problems.

Usually, in problems involving energy, the different forms of energy conversion are equated as the sum of the total energy within the system which is also conserved. With this parameter, it would be possible to solve specific values such as velocity, displacement, or any other parameter that the problem may require. Therefore, the law is quite useful for simple questions that have direct energy values, and their relations are present within a closed system.

FLUIDS

Matter exists in four key states: solid, liquid, gas, and plasma. Since solids have a defined shape, liquids and gases are considered fluids because they yield to shearing force and don't have a fixed shape, while solids resist and don't change their structure. Atoms in solids are very close to one another and have strong, attractive forces that don't allow them to change easily.

On the other hand, liquids deform easily because their atoms are free to move and slide amongst each other. This property is similar to atoms in gases; as, these are separated by great distances. The forces that hold gases together are fragile. Generally, both gases and liquids are referred to as fluids because of their similar characteristics, and the distinction between them is only made if they act differently.

PRESSURE

Imagine trying to hammer a bowling pin into the wall. Nothing would happen. Perhaps you might make a small dent, but, overall, the wall would remain the same. On the other hand, if you were to hammer a nail with the same amount of force, this would likely penetrate the wall. Hence, the magnitude isn't the only thing that matters; one also must know how the force is distributed on the surface of impact. For the nail, the force between the wall and the nail is concentrated on a tiny area on the tip of the nail. But, the area touching the wall is much larger for the bowling pin, making the force much less concentrated.

To elaborate on this concept, physicists used the idea of pressure. Pressure is defined to be the amount of force exerted per unit of area. To increase pressure, one would have to increase force or decrease area. This is why people are safe lying on beds of nails. Since the bed of nails is large enough to offset the force, this causes a minimal amount of pressure.

Both solids and fluids exert pressure. Imagine being submerged in some deep water. The water above would be pushing down on you because of the force of gravity, therefore exerting pressure on you. As you increase your depth, there will be more water above you to increase the pressure.

Similar to liquids, the gas above you can also exert pressure. Hence, the amount of air in our atmosphere is substantially above you and is continuously pressing on you. The reason you don't notice is because atmospheric pressure is always the same. Yet, when we fly or go underwater, we notice a change in pressure either above or below the usual amount. This large amount of atmospheric pressure doesn't harm the human body because our bodies are designed to withstand this amount of air pressure.

Usually, most gauges and monitoring equipment don't necessarily measure atmospheric pressure because that remains constant on Earth. Instead, this technology used gauge pressure. Gauge pressure is measured relative to atmospheric pressure. Additionally, the value of total pressure is known as absolute pressure. Absolute pressure, on the other hand, measures pressure relative to a complete vacuum. So, the absolute pressure can never be negative because a perfect vacuum is considered to contain zero pressure. There is a simple relationship between atmospheric, gauge, and absolute pressure: absolute pressure is equivalent to gauge pressure summed with atmospheric pressure. This makes logical sense because if you wanted to get the total pressure in a system, you would add the change in pressure from the original set amount. We will continue discussing pressure in further detail later in this chapter.

DENSITY

Density is the mass of a unit of volume for a material substance and is commonly expressed in CGS units of grams/cm^3, and the SI units are kg/m^3. Density offers a means of obtaining the mass of a body from its volume. Additionally, density is a property of matter such as mass or color. This also makes it a scalar quantity and not a vector quantity because it only has magnitude; there is no direction associated with density.

PASCAL'S LAW

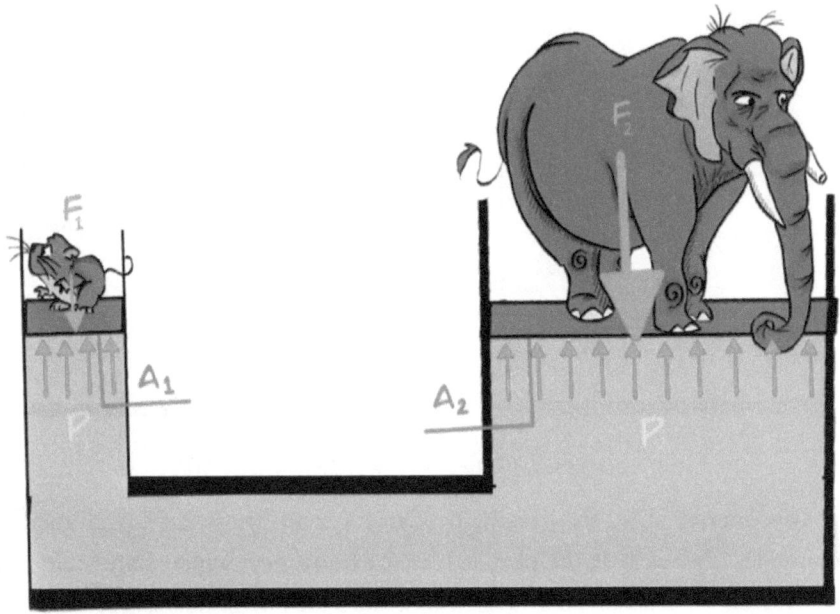

FIGURE 27: PRESSURE IS UNIFORM ON ALL SURFACES

Named after French scientist Blaise Pascal, Pascal's Law states, in a fluid at rest within a closed container, a pressure change in part of the container is transmitted equally to every portion of the fluid and the container's walls. In other words, in a hydraulic system, some pressure exerted on one piston produces an equal increase in pressure on another piston in the system.

Another aspect of Pascal's law is that the pressure at a point in a fluid at rest is the same in all directions. Simply speaking, this means that in a water balloon, the amount of pressure in one area of the water balloon would be equal to the pressure in another. This principle is why balloons filled with water or air are not lopsided when they are at rest. There is an equal amount of air or water pressure in all parts of the balloon, which leads to it expanding equally.

ARCHIMEDES' PRINCIPLE

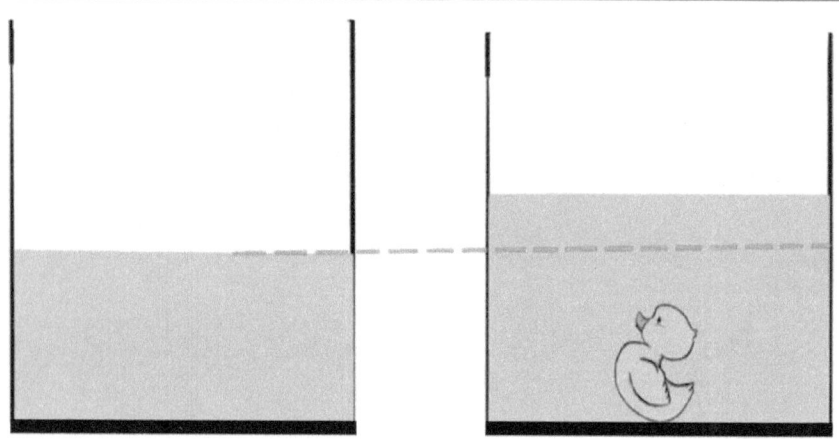

FIGURE 28: DISPLACED WATER VOLUME

Discovered by the ancient Greek mathematician Archimedes, Archimedes' Principle is the physical law of buoyancy, which states that any body entirely or partially submerged in a fluid (gas or liquid) at rest is acted upon by a force directed upwards. This force's magnitudes are equal to the weight of the fluid that is displaced by the body. In an object floating, the buoyant force has an equivalent magnitude to the floating object's weight in the opposite direction. The result is the object neither rising or sinking, but only floating.

A standard demonstration of Archimedes' Principle would be by placing a weight in a container of water. The weight of water that rises due to the introduction of the weight, or is displaced, is equivalent to the weight loss of the object submerged (the weight loss would be equal to the weight of water with the same volume as the object). Hence, if more mass is added to this weight, it would displace more water, showing the increased mass, increases the weight loss for the submerged body, and the larger buoyant force.

BUOYANT FORCE

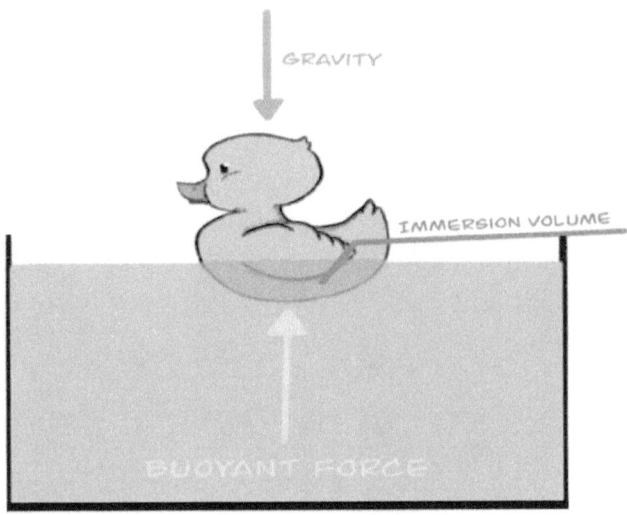

FIGURE 29: WEIGHT OF DISPLACED WATER IS EQUAL TO WEIGHT OF THE DUCK OR BUOYANT FORCE

Have you ever dropped something to the bottom of the pool and tried to swim down to get it? When swimming, you may notice that the water tries to push you back up to the surface. This upward force, which affects objects submerged in fluids, is known as the buoyant force.

The reason fluids exert an upward force on submerged objects has to do with the differences in pressure between the submerged object's bottom and top. If you were to drop a can into the pool, the pressure increases as you go deeper into the water. The pressure exerted downward on the top of the can will be less than the force exerted upward on the bottom of the can. Simply put, the buoyant force is an outcome of the unavoidable fact that the bottom of an object is always deeper than that top, resulting in unequal amounts of force and pressure.

FLUID CHARACTERISTICS

Fluids in physics have a couple of essential characteristics that help define the basis of fluid dynamics. One of the most crucial attributes of fluids is compressibility. Unlike solids, liquids deform easily and don't spring back to their original shapes since the atoms are free to slide around. Additionally, if a liquid is put into a container with no lid, it remains within the box without changing volume; it is important to note, liquids also resist compression, like solids.

On the other hand, gases are relatively easy to compress due to the large amounts of space between gas atoms and molecules. When placed in an open container, gases can escape; this is the primary distinction between liquids and gases in terms of compressibility. Generally, both gases and liquids are referred to as fluids, and the difference is only made when they behave differently.

BERNOULLI'S PRINCIPLE

Bernoulli's equation is a mathematical representation of Bernoulli's principle, a seemingly counterintuitive statement describing how fluids' speed relates to the fluid's pressure. Directly Bernoulli's principle states that, within a horizontal flow of fluid, higher fluid speed points will have less pressure than points of slower fluid speed. This statement may sound complicated at first, but let's break it down.

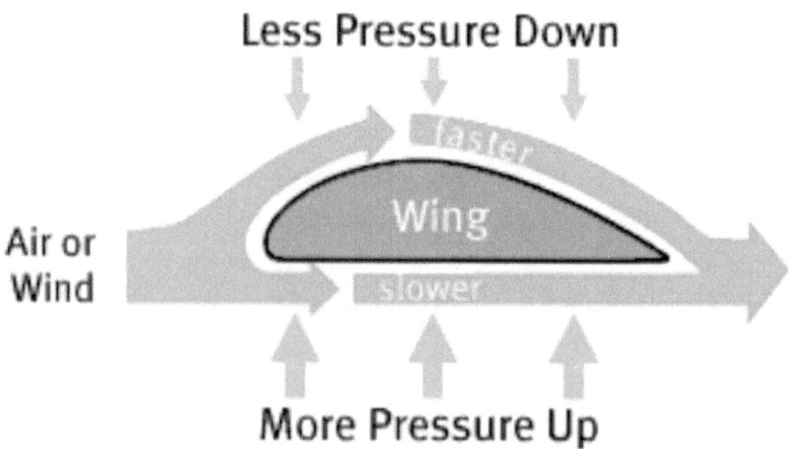

FIGURE 30: AIR PRESSURES ON AN AEROFOIL

A quick note, Bernoulli's principle focuses on the horizontal flow of a liquid; this is because, in a vertical flow, one has to consider gravitational potential energy (due to the change of height). Bernoulli's equation is slightly inaccurate in the vertical direction, even if Bernoulli's principle can be made more flexible to account for gravitational potential energy.

Simplifying Bernoulli's principle explains that within a horizontal pipe containing water whose diameter changes, the regions where water is moving fast will be under less pressure than areas with the water moving slowly. This seems like common sense since it's known that high speed results in decreased pressure. For the same reason, water pipes burst when the water moves through the tube at low speeds.

Incompressible fluids speed up when they reach a narrow area to maintain a constant volume flow rate. This principle is used in designing airplane wings to produce the lift necessary. The difference in pressure on the wings of the plane produces lift.

WAVES

The entirety of the universe is filled with vibrations or oscillations. All matter oscillates back and forth. These oscillations propagate as waves (e.g., ripples in the water) with energy, and at times information (e.g., radio AM/FM). All waves have specific common properties, such as periodic oscillations of the particles, also known as simple harmonic motion. Simple harmonic motion occurs when an object's oscillatory motion is proportional to the amount of displacement (e.g., the movement of a pendulum). Mathematically, these individual oscillations can be described using a sinusoidal trigonometric equation (e.g., the graphs of cosine, sine, etc.).

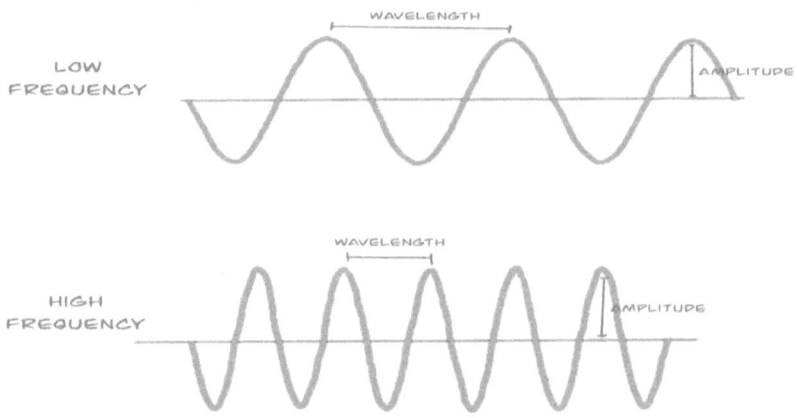

FIGURE 31: FREQUENCY AND PROPERTIES OF WAVES

Waves are known as an energy transfer phenomenon. Also, along with transferring energy, waves can also share information (e.g., wireless communication). In general, everything in the universe is, inherently, a wave. As we will further discuss, all atomic particles behave like both waves and particles. Usually, waves transfer energy through the use of disturbance. These disturbances can happen in any matter or space. For example, if an individual were to bang a drum, the mechanical energy is transferred as sound waves, vibrations, and heat. Even a supernova explosion continuously generates electromagnetic waves, scalar waves, gravitational waves, and several others, such as space radiation (which we will cover further in this book).

Over time, all waves fade after traveling a certain amount of distance. Water waves provide the most visual example. They become smaller as they travel further and further away from the source. This fading or attenuation of amplitude occurs because ripples expand from the origin in concentric spheres (in 3D) or concentric circles (in 2D). This expansion also decreases the energy transmitted or the intensity. As the energy is distributed over a large sphere (or ring) with increasing distance, the intensity decreases. So, the wave's intensity inversely proportional to the square of the distance traveled. Thus, the wave's intensity is inversely proportional to the square of the distance traveled. An excellent way to imagine and understand this is to listen to someone whisper standing a couple of feet away; it would be quite tricky. Now, imagine the same person whispering in your ear. It would be much easier to hear. This is simply because of the sound's intensity (volume) increases as the object reaches our ear. This relationship between intensity and distance is the same for all types of waves (inverse square law).

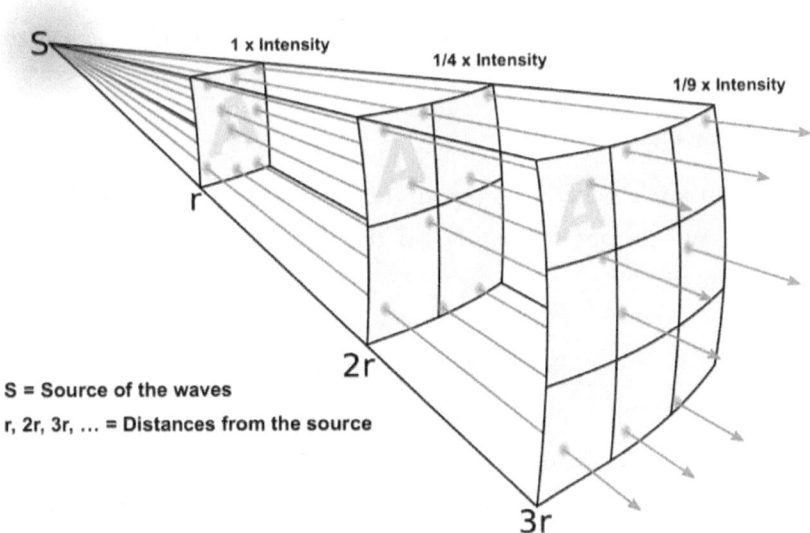

FIGURE 32: WAVE INTENSITY REDUCTION WITH DISTANCE [8]

In nature, an individual would observe five different types of waves:

1. Mechanical Waves

   Mechanical waves are quite prevalent throughout our world today. Examples include sound waves, water waves, ocean waves, seismic waves, etc.

2. Electromagnetic Waves

   Sunlight, X-Rays, WiFi wireless, radio transmissions, etc., are all examples of electromagnetic waves. Electromagnetic waves, specifically coming from the Sun, also contain tiny particles called photons.

3. Scalar Waves

   Additionally, the Sun and other interstellar objects emit space and neutrino radiation (a Neutrino is an atomic particle with almost zero mass and zero charges, we will discuss this further) and scalar waves which are longitudinal waves.

4. Matter Waves

   Matter waves are waves generated by atomic particles like electrons, protons, neutrons, etc.

5. Gravitational Waves

   Einstein's general theory of relativity shows that cosmic entities would disrupt space-time so that "waves" propagate around the source.

WAVE PARAMETERS

The five different types of waves mentioned above propagate as either transverse or longitudinal waves due to the disturbance amongst particles. There are six parameters associated with every wave: phase velocity, frequency, amplitude, wavelength, and period of oscillation. Phase velocity is the speed of wave propagation. Frequency is measured in oscillations per second. For example, the human ear has a frequency range from 20 Hz to 20,000 Hz (Hz, also known as Hertz, is the unit used to measure frequency in cycles per second). A wave's amplitude is the maximum amount of displacement or distance that a particle in a wave moves from the rest position. Wavelength is, quite simply, the length of a wave for one full oscillation. The period of oscillation is the time it takes for one oscillation.

There are two main formulas associated with the six parameters mentioned above. No matter the wave, these two formulae are consistently witnessed in nature:

Speed of the wave (Phase velocity) = Frequency X Wavelength

Period of oscillation = 1 / Frequency

TRANSVERSE WAVES

Transverse waves are waves whose oscillations are perpendicular to the direction of propagation. Electromagnetic waves, ocean waves, and water waves are all transverse waves. For instance, ripples in a lake associated with throwing a stone happen horizontally, but the individual water particles move up and down.

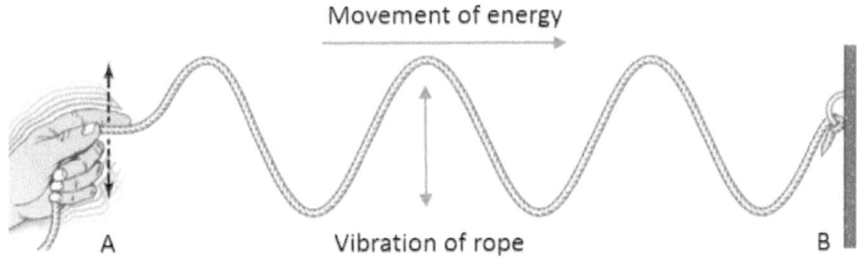

FIGURE 33: ROPE'S TRANSVERSE WAVE OSCILLATIONS [9]

Polarization is the relationship between the direction of oscillation and the propagation of the wave. Hence, transverse waves are said to be vertically polarized. Another form of a transverse wave is a standing wave, which can be created by tying a rope to a rigid object, such as a wall (we see people exercising this way in the gym at times). The exciting thing about a standing wave is that the net phase velocity is zero (also known as the wave's total velocity). In simpler terms, how fast the wave is moving.

LONGITUDINAL WAVES

Longitudinal waves are waves whose oscillations are parallel to the direction of the wave. Sound waves, scalar waves, plasma waves, and matter waves propagate this way.

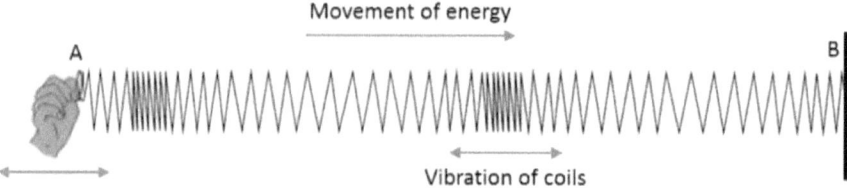

FIGURE 34: LONG SPRING'S LONGITUDINAL WAVE OSCILLATIONS [9]

For example, when a spring vibrates, the particles associated with these waves oscillate parallel to the spring's vibration.

GENERAL PROPERTIES OF WAVES

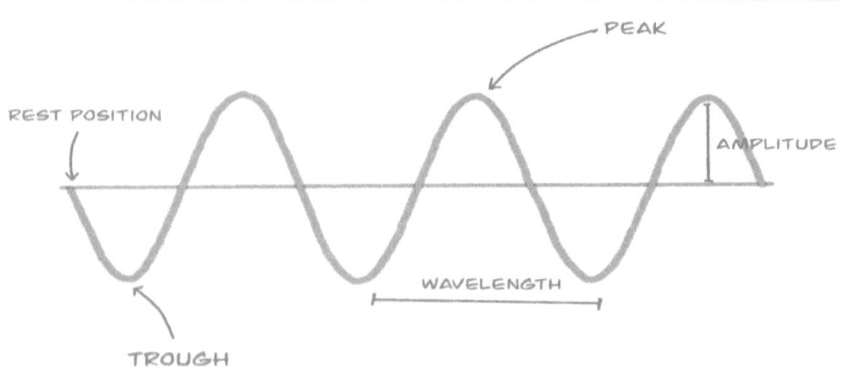

FIGURE 35: MEASURING QUANTITIES OF A WAVE

All waves share specific general properties—the first being wavelength, which is the distance between neighboring crests and troughs. A crest or peak is the highest point in a wave, and the trough is the lowest point. Waves also have a speed, which is the rate at which the crests or troughs move forward. After determining the speed and wavelength, one ends up with a period, which is the time between passing peaks or troughs. Specifically, the period is wavelength over velocity or speed. Frequency is the inverse of the period, which is also the number of crests or troughs that pass per unit of time.

REFLECTION AND REFRACTION

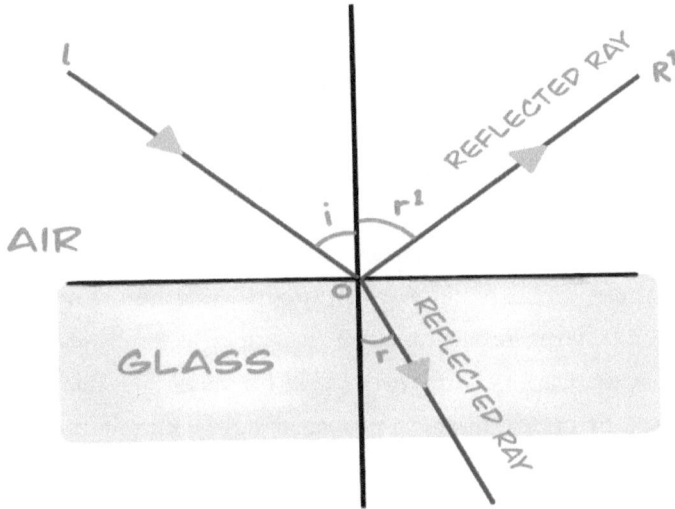

FIGURE 36: REFLECTION AND REFRACTION OF LIGHT

All waves can reflect and refract from different surfaces. The most famous wave when addressing this property would be light. We commonly see our reflection in a mirror every day; this would not be possible unless the light reflected from the surface and depicted your image. Along with this, I'm sure everyone has seen a rainbow. That is what occurs when light refracts in the atmosphere. This property occurs within standing ways as well; the wave's propagation is reflected at the wall through the rope.

Reflection occurs when the wave encounters a new medium through which it's unable to propagate. On the other hand, refraction happens when the wave passes through a medium where the speed of the wave changes. Additionally, when a wave strikes another medium at an inclination, the wave propagation changes direction because of the change in speed or refraction. I understand that this may seem confusing. So let's take an example.

In a rainbow, the sunlight entering the clouds gets refracted into seven different components because white light changes its propagation speed and bends different amounts based on the sunlight's inclination. Additionally, some portion of the wave gets absorbed by the medium, and the others get reflected or refracted out.

TOTAL INTERNAL REFLECTION

As elaborated on previously, refraction happens when a wave strikes a medium at a specific angle. During this process, the wave propagation direction bends at an angle different from the original angle of the incident (the angle at which the wave hits the surface is called the angle of incidence). Suppose that the angle of incidence is adjusted so that the entire wave turns back from the source, known as total internal reflection. And the angle at which the total internal reflection occurs is defined as the critical angle. When this happens, the wave gets trapped within the medium—this is a common property used by optical fibers to propagate light with minimal loss in long distances.

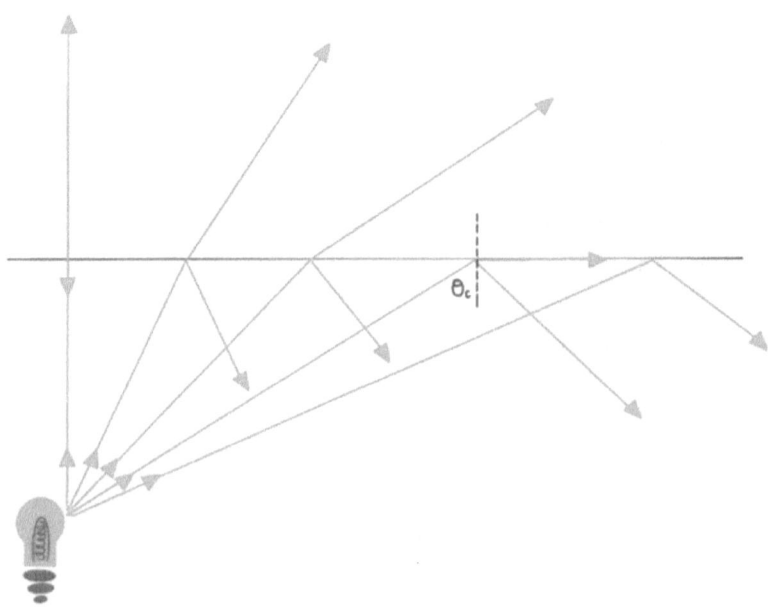

FIGURE 37: PROCESS OF TOTAL INTERNAL REFLECTION

INTERFERENCE

Interference is an event that occurs when two or more waves move through the same medium at the same time. When someone hears two or more sounds simultaneously, this is a clear everyday example of interference. When these two sounds are played together, both waves get mixed, and the individual hears the combined wave. Interference is also used for modulating the waves for long-distance transmission. The long-distance transmission we are familiar with is AM (Amplitude Modulation) and FM (Frequency Modulation).

FIGURE 38: INTERFERENCE OF WATER WAVES

DIFFRACTION

Diffraction refers to a change in the direction of waves as they pass through an opening around a barrier in their path. This is the same reason that water waves can travel around corners, obstacles, and gaps. Suppose you had a tank and placed small barriers into that region and dropped a stone into this tank. You will notice that the waves pass around the obstacle and the barrier into the areas behind it. This, in turn, results in the water behind the barrier being disturbed. The diffraction amount is directly proportional to the wavelength (as wavelength increases, diffraction increases, and vice versa).

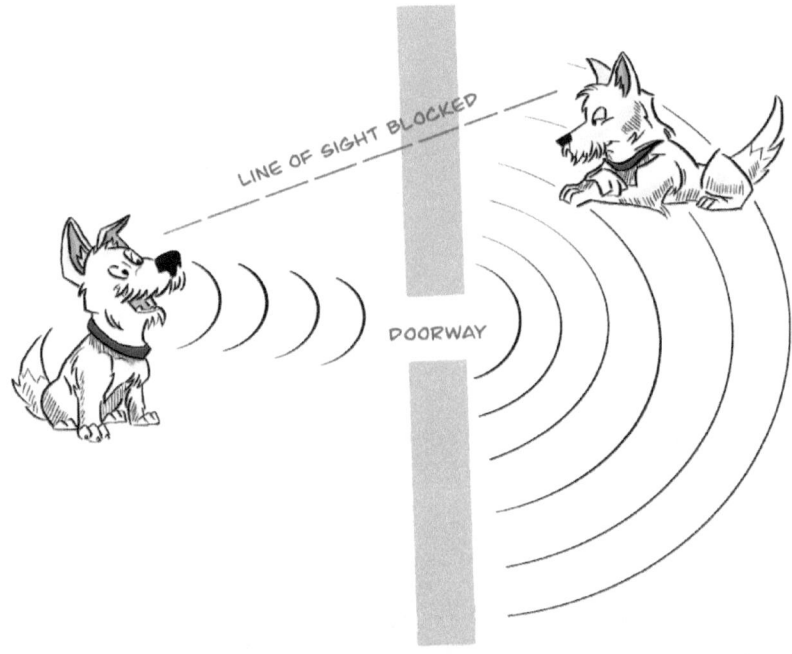

FIGURE 39: DIFFRACTION OF SOUND WAVES

Though diffraction occurs in water waves, it's also quite prevalent in both sound and light waves. This diffraction results in us hearing people in other rooms, and echolocation works with the same principle. Diffraction is observed among light waves when light encounters either incredibly tiny passages or obstacles.

RESONANCE

All bodies in nature have multiple natural frequencies of vibration for resonance. For example, if one were to hold a thin metal rod and hit it against a hard object like a wall, the metal rod vibrates at a particular frequency. This frequency is its natural frequency of vibration. If this metal rod is placed in the path of a sound wave, which is also at its natural frequency, then the metal rod will start resonating with the sound waves. The amplitude of the rod's vibration will become extremely high since it's vibrating at its natural frequency. This phenomenon is called resonance.

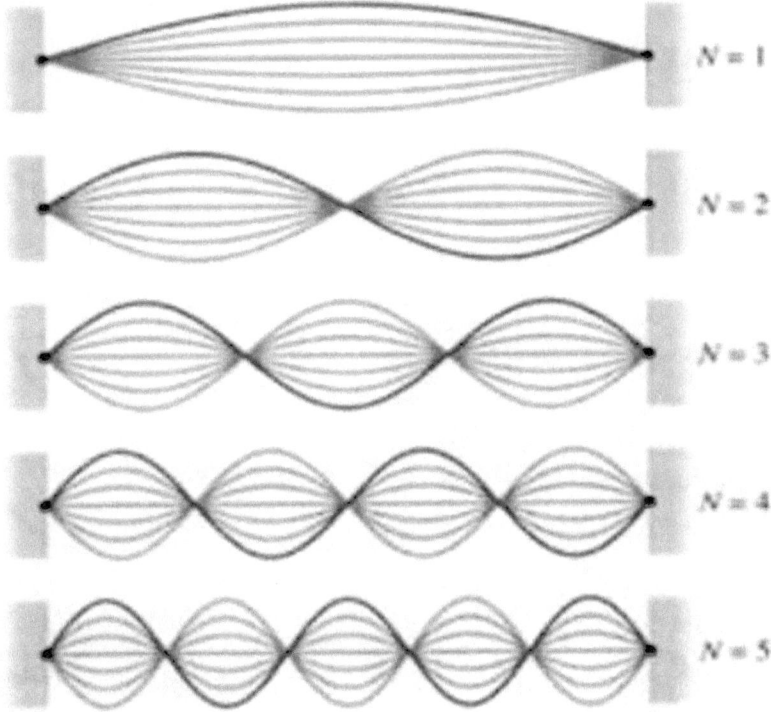

FIGURE 40: VARIOUS RESONANCE FREQUENCIES OF A STRING [10]

When resonance occurs, since the vibration's amplitude is very high, the object vibrating might even break up in some cases. One can reach the natural vibrational frequency in a multitude of ways. Any disturbance, for example, can cause the body to vibrate. Sound waves can also cause this disturbance, but there are multiple other ways for it to occur in nature. This phenomenon happens in all kinds of waves, such as sound waves, electromagnetic waves, scalar waves, mechanical vibrations, etc. Additionally, this is true for even astronomical bodies, which also resonate with their natural frequencies.

Resonance can also happen between multiple objects. Suppose a body is vibrating at its natural frequency, and another body enters the system. It has been noticed that both bodies start resonating with a frequency suitable to both. In mechanical vibrations, air molecules transfer vibrations from one body to another so that both bodies start vibrating with the same frequency.

As mentioned earlier, resonance happens with all types of waves and radiations, including astronomical bodies. However, since astronomical bodies are immense, vibrations with the resonance frequency will be very difficult to measure.

THE DOPPLER EFFECT

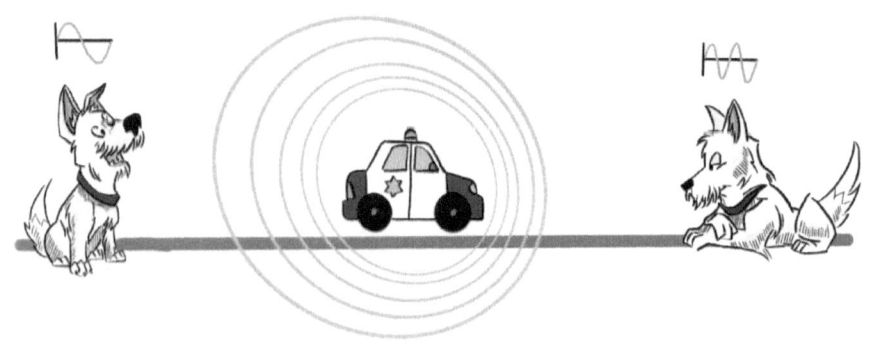

FIGURE 41: DOPPLER EFFECT WITH A MOVING VEHICLE

When a source generating waves is in relative motion concerning a receiving station, there is an observed change of frequency in the receiving station. This phenomenon is known as the Doppler Effect, and it occurs with all types of waves. A clear everyday example of the Doppler Effect is a siren on a police vehicle. When the police car is driving towards you, a different sound is heard than when the car drives away from you. The siren sounds like a higher frequency when it comes towards you and lowers as it moves away. The Doppler Effect is also witnessed in stars, with the light that they emit. The electromagnetic frequencies (e.g., light) change depending on the star's motion and direction with respect to Earth. When a star moves away from us, we see a redshift, and when it moves towards us, we view a blueshift.

STANDING WAVES

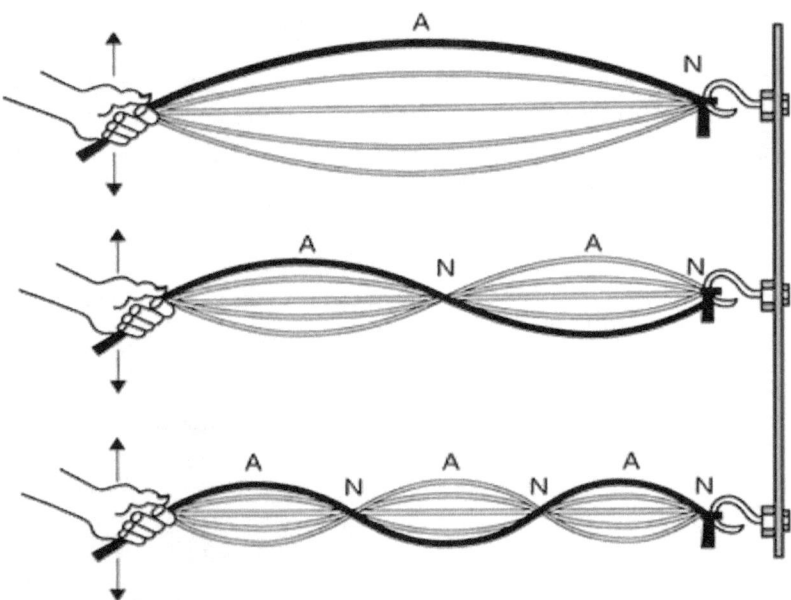

FIGURE 42: STANDING WAVE HARMONICS WITH A ROPE

A wave that travels down a rope gets reflected at the rope's end. If the end of the string is free, the wave returns right side up. If the end is fixed, then the wave will be inverted. For a rope with two fixed ends, another wave traveling down the cord will interfere with the reflected wave. Depending on the frequency, this produces standing waves where the nodes and antinodes remain in the same places over time. Each successive harmonic has an additional node and antinode; this means that the first harmonic has one bump, for the second, there would be two bumps, and so forth, where a string has an infinite amount of resonating frequencies.

ENERGY AND POWER OF WAVES

All waves carry energy; this is commonly observed through natural disasters such as earthquakes and water waves. The amount of energy in a wave is related to its amplitude and frequency. Note, the total amount of energy in a mechanical wave is the sum of its kinetic energy and potential energy. Loud sounds have high-pressure amplitudes; hence, the larger the amplitude, the larger the change in potential energy.

High-frequency waves will deliver more "packets of energy" per unit of time than a low-frequency wave. So if two waves have equal amplitudes, but one has a higher frequency, it would also carry more energy. Additionally, in mechanical waves, the rate of energy transfer is proportional to both the square of the amplitude and the square of the frequency.

The power of a sinusoidal mechanical wave is the average rate of energy transfer associated with the wave as it passes a point. This value is found by taking the total energy with a wave divided by the time it takes to transfer the energy. Additionally, in a mechanical wave, the power is proportional to the square of a wave's amplitude and the square of the frequency.

WATER WAVES

We are familiar with water waves in the oceans, lakes, and rivers. Water waves are the most complex type of waves as they contain a mixture of ripples with different sizes, frequencies, speeds, and durations. Also, when the water depth becomes shallow, the wave slows down and becomes short-lived. There are many types of waves and various effects within water (some of which will be explained below).

HELMHOLTZ RESONANCE

As a child, everyone has probably tried to make a sound by blowing into an empty bottle. This sound is a product of a resonance phenomenon known as Helmholtz resonance. The resonance occurs because the cavity's body is filled with air, whose volume is that of the bottle. The neck of the bottle also contains some air that has a mass. When one blows into the open channel, the air inside the neck is pushed down into the body, leading to a compression of the air inside. This compression imbalance of the restoring force expands the air, and the amount of air in the neck that was initially there is pushed upwards. Since the mass has inertia, the mass passes through equilibrium and is compressed again. This cycle repeats itself and is viewed as simple harmonic motion, thereby producing resonance. The resonance of this frequency can be determined using: the volume of the bottle, the correctional area of the neck, and the neck's length.

FIGURE 43: HELMHOLTZ RESONANCE WITH EMPTY BOTTLE

HARBOUR RESONANCE

When a harbor is connected to an open sea with a narrow water passage, a resonance occurs. The opening passage behaves like a small bottle opening (as in Helmholtz resonance). This sort of resonance in open water is called the harbor resonance. This resonance causes the wave sizes to be quite large, which means that small boats need to be careful.

SEICHE AND TSUNAMI

Seiches and tsunamis are commonly grouped; yet, they are two entirely different events. A standing wave oscillating within a body of water is called a seiche. You've probably witnessed a small-scale seiche in your bathtub or swimming pool if the water sloshes back and forth. On a much larger scale, the same phenomenon occurs in large bodies of water. Usually, seiches commonly occur in semi or fully-enclosed areas, thereby not making them very common in the open ocean.

FIGURE 44: TSUNAMI WAVE AFTER AN EARTHQUAKE IN THE OCEAN
[11]

In these large bodies of water, seiches are caused when strong winds and rapid fluctuations in atmospheric pressure push the liquid from one end of a lake to another. When the wind stops, the water rebounds and continues to oscillate back and forth for incredibly long periods of time. Tsunamis may also cause seiches along ocean shelves or harbors that could do a lot of damage.

On the other hand, tsunamis are giant waves caused by earthquakes or volcanic eruptions occurring underwater. Usually, in the depths of the ocean, tsunami waves don't dramatically increase in height; however, as they travel inland, they can build up as the depths of the ocean decreases. A tsunami's speed can also increase as it reaches inland since the water's depths affect the wave's speed. These waves can travel as fast as jet planes (usually over deep waters) and ultimately slow down when reaching shallower areas. As shown, two completely different circumstances cause seiches and tsunamis, thereby making them completely different geological events.

TIDES

The creation of tides has consistently dumbfounded many scientific thinkers for centuries. Today it's commonly known that the gravitational pulls between the Earth, moon, and Sun dictate the tides, with the moon having the most influence.

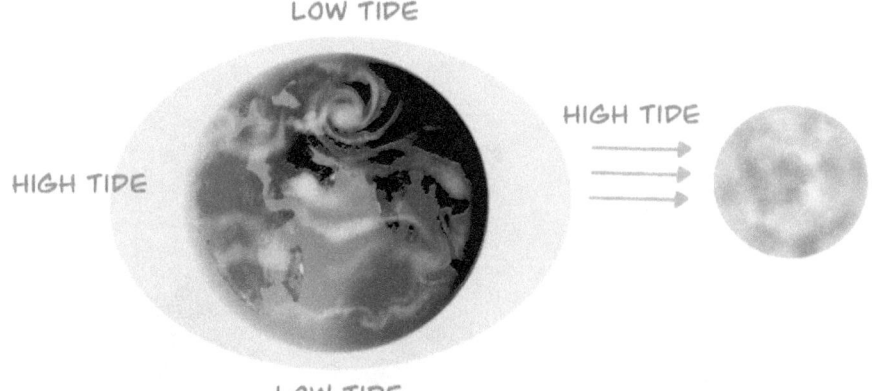

FIGURE 45: EARTH'S OCEAN TIDES DUE TO MOON'S GRAVITATIONAL ATTRACTION

The moon's gravitational pull is powerful enough to cause the oceans to bulge. But, if no other forces were present, shores would experience high tide on the beaches that face the moon. Due to inertia, the centrifugal force also ends up affecting the tides. As the moon orbits the Earth, the Earth moves in a very slight circle as well. This small movement is enough to cause a centrifugal force on the oceans; this results in oceans bulging on the opposite side facing the moon. Though the moon's pull is strong enough to attract oceans on the side of the Earth-facing it, it's not strong enough to overcome the centrifugal force on the opposite side of the Earth. As a result, oceans bulge twice, once when they're on the side closest to the moon and when they're on the side farthest from the moon.

Though the moon is a more prominent figure on the rise and fall of tides, the Sun also similarly affects the tides. "Solar tides" are due to the Sun's gravitational pull and are generally weaker than lunar tides. The Sun is around 27 million times more massive than the moon and 390 times father way which makes it about half as powerful as the moon. So, solar tides are considered variations of lunar tides. In general, tides can also be affected by a location's geography, leading to larger or smaller tides depending on the site.

ELASTICITY

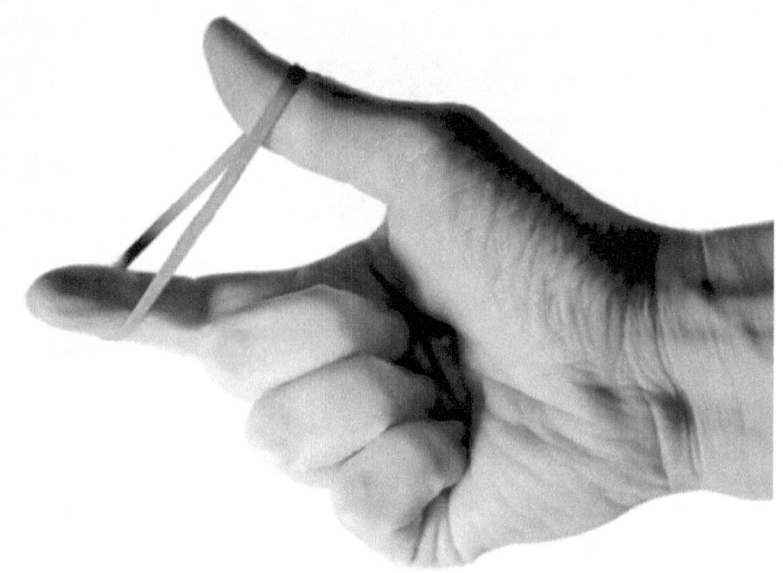

FIGURE 46: ELASTICITY OF A RUBBER BAND

In solids, the internal atoms and molecules that make up the substance are bound together using tiny springs or attractive bonds. These small springs create the solid's known shape, and since these forces are incredibly strong, a solid, generally, keeps its shape. In some instances, such as when a long rod is hit against a wall, it bends slightly due to the force applied. The reason for this is because hitting stretches these bonds, resulting in the deformation. Yet, immediately afterward, the rod regains its shape due to these attractive forces pulling the atoms back into their original position. This property of solids is known as elasticity, and different solids have different levels of elasticity depending on the strength of their internal bonds.

Elasticity causes vibrations in objects, and all items in the universe oscillate or vibrate, including atoms and molecules. Make a note that the internal bonds in liquids and gases aren't as strong as solids. Hence their shape is not maintained.

STRESS AND STRAIN

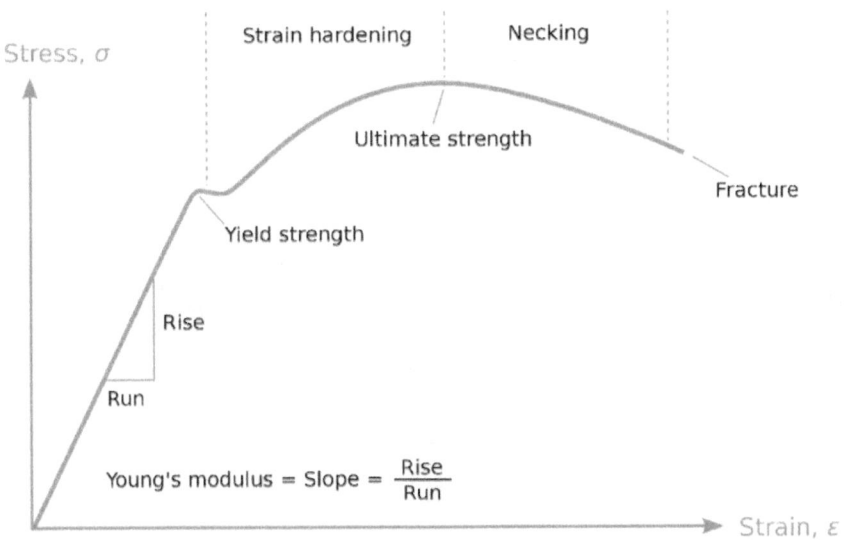

FIGURE 47: YOUNG'S GRAPH OF ELASTICITY [12]

As stated previously, the most common form of elastic potential energy is that of a spring. Spring is an object that can be deformed by force and returns to its original shape after the force is removed (like most elastic objects). Though there are various types of springs, the most familiar would be the simple metal coil.

When a force is applied to a material (such as a spring), the material stretches or compresses to respond to the force. The extent of the deformation produced as the material responds to the force's stress is known as strain. Strain is measured as a ratio of the difference in length to the original size. Every material responds differently to stress, which depends on the tightness of the chemical bonds within the material.

When studying springs and elasticity in the 17th century, physicist Robert Hooke noticed a linear relationship between the stress and strain of many different materials. Within certain limits, the force required to stretch an object like a metal spring is directly proportional to the spring extension. This concept is known as Hooke's law.

ELASTIC POTENTIAL ENERGY

Elastic potential energy is the energy stored due to applying force to deform an elastic object. This energy is stored in the item until the force is removed, then the object returns to its original shape, doing work in the process. The most common form of elastic potential energy is either compressing or stretching a spring.

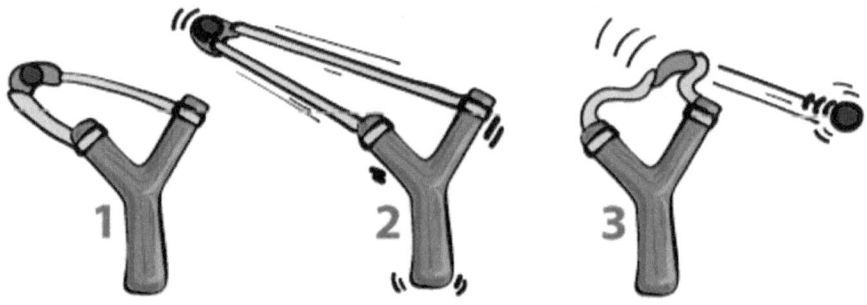

FIGURE 48: CONVERSION OF ELASTIC POTENTIAL ENERGY TO KINETIC ENERGY [13]

An object, such as a spring designed to store elastic potential energy, will have a high elastic limit. Yet, all elastic objects have a limit to the amount of load that they can sustain. If an item is deformed beyond its elastic limit, it will not return to its original shape; this is known as plastic deformation. Elastic deformation is when the force is removed, the object returns to its original form, and this deformation is non-permanent and reversible.

POISSON'S RATIO

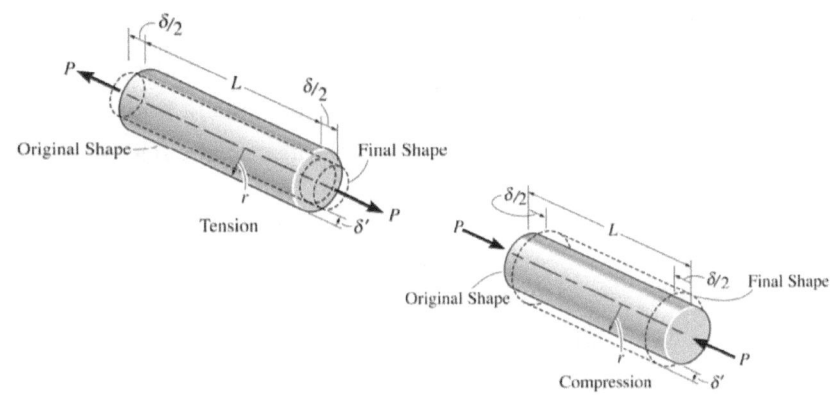

FIGURE 49: POISSON'S RATIO BASED CHANGE IN SHAPE [14]

Let's start with a quick example if someone were to take a rubber band and stretch it, as its length increases the band, itself, will get thinner. Poisson's ratio tells us how much thinner the band will get when a force is applied. The key concept is that if a force is applied to a material (such as an iron rod) in one direction, it will deform in the lateral direction. For instance, if you were to squeeze a material, it would get taller and thinner. On the other hand, if one were to stretch the same material, it would get thinner and longer. Poisson's ratio tells us how much a material will deform in the lateral directions.

It's important to note that Poisson's ratio only applies to isotropic materials. Isotropic materials are objects that have the same properties in all directions. It's also assumed that the objects are deforming within the elastic region; things get complicated when plastic deformation is involved. Put simply, Poisson's Ratio is a dimensionless material property, which informs us how much a given material will contract in the lateral directions when one pulls on it in the longitudinal direction.

All objects theoretically have a Poisson's Ratio between -1 and 0.5. Objects whose ratio is between 0 and 0.5 become thinner as they are pulled in the longitudinal direction. Objects whose rating is precisely zero behave interestingly. When compressed, these objects remain the same size. A simple example of such an item is a cork. Due to the Poisson Ratio of a cork being zero, it can be easily inserted into a wine bottle, while another material would contract. Finally, objects with a Poisson's ratio from 0 to -1 expand in the lateral direction when stretched. This idea may seem counterintuitive; however, these are mostly engineered materials, like unique foams.

ELASTIC PROPERTIES OF MATERIALS

It is commonly assumed that rigid bodies don't undergo any deformations due to applied forces. On the contrary, real objects deform once forces are applied. In physics, the description of such deformation within objects is known as stress and strain. Stress is defined as force per unit area. For example, when forces pull on an object and cause it to stretch, this is known as tensile stress.

On the other hand, strain is the measure of how much an object is stretched or deformed, specifically a change in the volume or shape of an item. Specific forms of strain are reversible while others are not, and this depends on the type of stress that resulted in the deformation. For example, an elastic strain is reversible; but, shear strain (when scissors cut thin material) is not.

If stress and strain are linearly proportional in a specific material, then that material is known to obey Hooke's Law. Specifically, Hooke's Law only holds true for a range of stresses referred to as the elastic region. Outside of this, the stress-strain relationships become non-linear until the object breaks. Within Hooke's Law, if a body is initially deformed due to the application of stress but returns to normal after the stress is removed, then the body is known as an elastic body. An idealized spring is an example of an elastic body.

Additionally, there are several different types of stress:

1.) Tensile Stress: tension induced to a body when the body is subjected to equal and opposite "stretching" forces.

2.) Compressive Stress: stress-induced to a body when the body is compressed.

3.) Shear Stress: when the body is induced to two equal and opposite forces acting and subjecting the body to a torque.

Conversely, several different types of strain:

1.) Tensile strain: force leads to a decrease in area and an increase in the body's length.

2.) Compressive strain: force leads to an increase in area and a decrease in the body's length.

3.) Shear strain: there is a form of angular distortion of the body due to the force (or induced torque due to the force).

VIBRATIONS AND SOUND

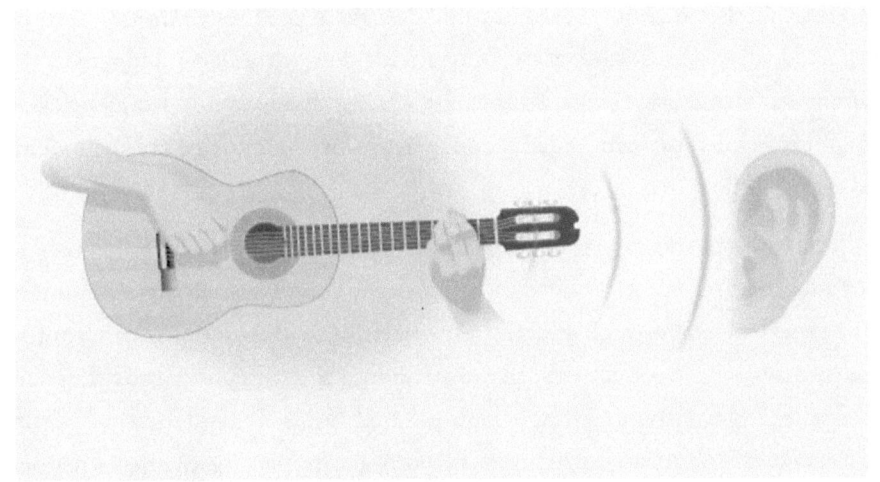

FIGURE 50: SOUND WAVES WITH VIBRATING STRINGS

All matter: atomic particles, atoms, molecules, human DNA (Deoxyribonucleic Acid), cells, planets, stars, etc., vibrate. There is no single particle in the Universe that is entirely at rest; vibration is an essential property of all matter. These vibrations allow particles to interact with others and exchange tiny amounts of energy that create all the physical phenomena in the Universe.

Sound waves are longitudinal, mechanical waves; therefore, they need a mechanical medium such as solids, liquids, or gasses to propagate. It is for this reason that nobody can hear you scream in space. Since space is a vacuum, devoid of mechanical mediums, sound waves cannot travel, leading to absolute silence.

HUMAN AUDIO RANGE

Humans use the sound frequency in the range of 20 Hz to 20,000 Hz for daily communication, where the ear passes these sound vibrations to the brain as chemical signals. The hearing range of animals is mostly different compared to that of humans. Note that even though humans cannot hear higher frequencies, the human ear passes the sounds with frequencies more than 20,000 Hz to the brain. And consistently, it has been seen that the human brain understands these signals or frequencies when modulated in a certain way. Even though 20 Hz to 20,000 Hz is a decent range, as can be expected, there are sound wave frequencies that are much higher than the human ear's capacity. These sounds are known as ultrasonic frequencies. A small range of these ultrasonic frequencies is used to create unique sounds called subliminal messages.

Subliminal messaging is a common technique used to send signals to our subconscious mind by bypassing our conscious or hearing brain. Humans also experience high-pitch at high frequencies of sounds and low-pitch at low frequencies of sound. Musical instruments use the human ear's experiences to create various sound effects: high-pitched notes, low-pitched notes, the resonance of sound waves in an air column, harmonics, beats (beats happen when slightly different frequency sound waves are mixed), etc.

The speed of sound depends on the properties of the medium in which it is traveling (solid, liquid, or gas). Generally, the speed of sound is highest in solids and slowest in gases (in air, at 20°C, sound travels at 343 meters per second). Objects traveling below the speed of sound are known to travel at subsonic speed. Objects at the speed of sound travel at a sonic speed and objects above the sound of speed are traveling at supersonic speeds (e.g., supersonic fighter aircrafts). Other than supersonic fighter aircrafts, there are many applications of sound waves. Seismic prospecting teams use sound waves for finding oil, ships (sonar) to detect underwater objects, and submarines use these sound waves to track other submarines, etc.

PITCH

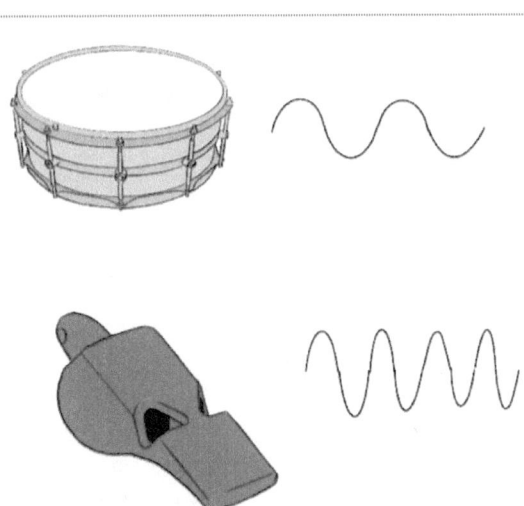

FIGURE 51: SOUND INSTRUMENTS WITH DIFFERENT TYPES OF PITCH

Sound waves are vibrations in a medium that can be deciphered by the human ear. The pitch of a sound is the description of its frequency. A high pitch corresponds to a high frequency, and a low frequency represents a low pitch. Amazingly, humans can distinguish minor changes in pitch (5000 distinct pitches) compared to changes in colors.

Individual sound waves played together to produce a pleasant sensation in the ear known as a consonant. These intervals are what create the basis of musical theory. For example, if two sounds whose frequencies make a 2:1 ratio are said to be separated by an octave, musically speaking. Additionally, these two waves sound good if played together.

The reason humans can differentiate pitches is that sound waves traveling through the air are longitudinal waves. These waves produce high-and-low pressure disturbances of air at a given frequency. These disturbances can be detected by the ear allowing humans to differentiate even minor changes in pitch.

HARMONICS OR OVERTONES

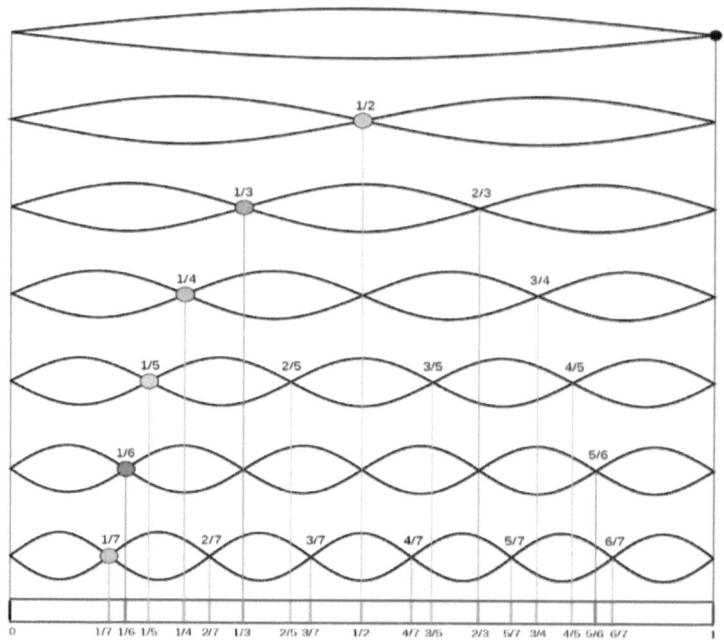

FIGURE 52: HARMONICS WITH WAVES [15]

These standing waves are especially crucial in the case of musical instruments. Suppose we tie a rope to a wall and shake it. Initially, we create a wave that looks like an oval. If we keep shaking harder, we will see more ovals. This phenomenon of forming ovals is called harmonics or overtones, and the frequencies at which these ovals form are known as resonant frequencies.

BEATS

When two sound waves of a similar frequency interfere with one another, they create periodic and repeating fluctuations, also known as beats. A beat pattern consists of a wave whose amplitude is changing at a consistent rate. During the formation of this pattern, two types of interference occur, constructive and destructive. Constructive interference is associated with a loud sound being heard, while during destructive interference, no sound is heard. Therefore, a beat pattern is consistent with a wave whose volume increases at a regular rate.

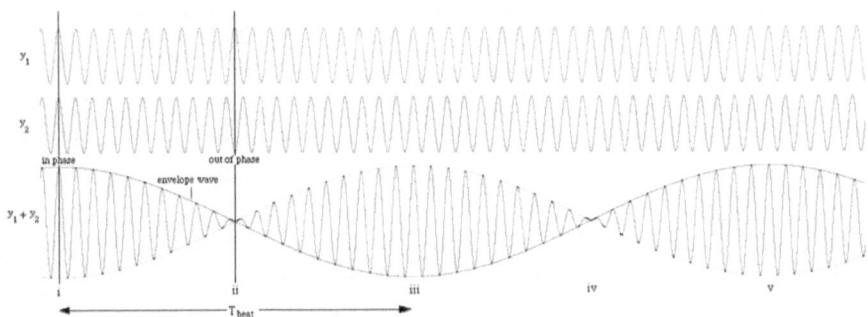

FIGURE 53: BEATS GENERATED WITH TWO WAVES

Now, moving to the beat frequency. As stated previously, the volume of a beat pattern increases and decreases at regular intervals. The beat frequency is the rate at which the volume oscillates between high and low values. This frequency is always equivalent to the difference between the frequencies of the two interfering waves. So, if two sound frequencies of 256 Hz and 257 Hz are played together, the beat frequency would be 1 Hz. The human ear is incredibly sensitive and can detect beats with frequencies of 7 Hz and below.

There are audio files created with different sound frequencies in the left and right ears. The difference between the value of these left and right audio frequencies will generally be less than 20 Hz. This audio is called "Binaural Beats," which allows our human ear to listen to sound frequencies below the lower range of 20 Hz. These lower range frequencies are useful for transferring brainwave frequencies for interacting with the brain, which we will learn later on.

INFRASOUND

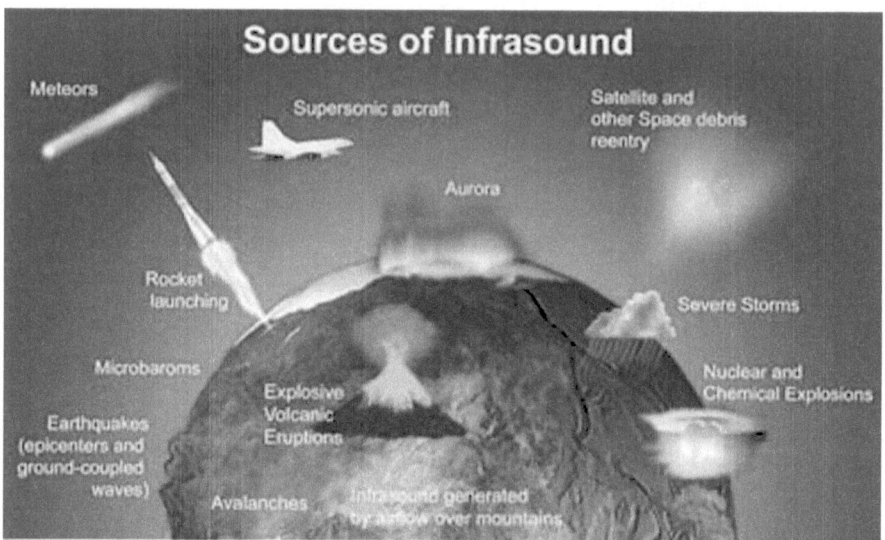

FIGURE 54: VARIOUS SOURCES OF INFRASOUND ON EARTH [16]

Frequencies below 20 Hz are known as infrasound. Nature produces many sounds (lightening, airwaves, etc.) in the infrasound range. Humans seem to have an emotional response to such sounds, and contrary to common belief, there have been circumstances where the human ear can hear sounds as low as 1.5 Hz. Most of our brainwave frequencies are below 20Hz.

ULTRASOUND

Ultrasound or sonography uses high-frequency sound waves to create images of the inside of the body. This technique is commonly used during pregnancy due to its safety, in comparison to other methods. The term ultrasound are sound waves that travel through soft fluids and tissue. Additionally, these sounds cannot be heard by humans. Higher frequencies provide better quality images because the skin and other tissue more readily absorb them, so they cannot penetrate lower frequency sounds. Though lower frequencies penetrate deeper, the image quality is inferior.

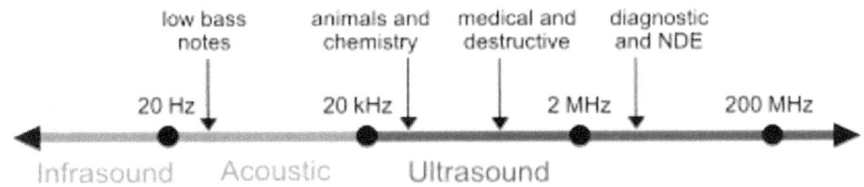

FIGURE 55: VARIOUS SOUND WAVE RANGES [17]

SEISMIC WAVES

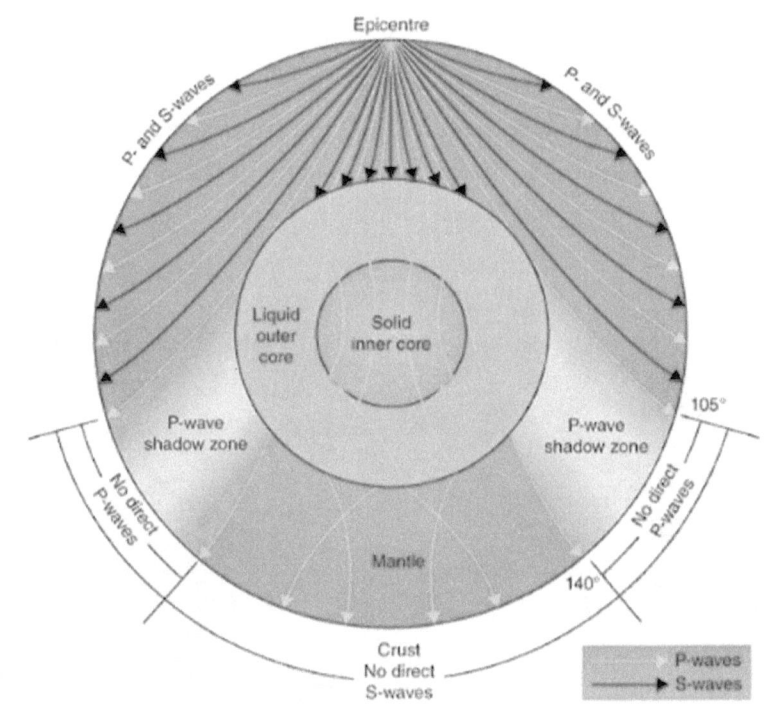

FIGURE 56: PROPAGATION OF P-WAVES AND S-WAVES AFTER AN EARTHQUAKE [18]

During earthquakes or explosions, vibrations are generated, which propagate within the Earth or along its surface. These vibrations are known as seismic waves. Earthquakes create four types of elastic waves. Two are known as surface waves that travel along the surface of the Earth. The other two are body waves, which travel inside the Earth.

P-WAVES

The first kind of body wave is known as primary waves (p-waves). During an earthquake, these waves "arrive" first. Additionally, these waves can move through solid rock and fluids, hence why they can travel so smoothly through the Earth's different layers. P-waves are also known as compression waves. Particles in a p-wave move in the same direction as the wave propagation, thereby making these longitudinal waves.

S-WAVES

The next type of body wave is the s-wave or secondary waves. These waves arrive after p-waves and are commonly witnessed during aftershocks in an earthquake. These waves, unlike p-waves, are transverse and are unable to travel through liquids. The particles in S-waves travel perpendicular to the direction of wave propagation (the primary definition of a transverse wave).

BRAINWAVES

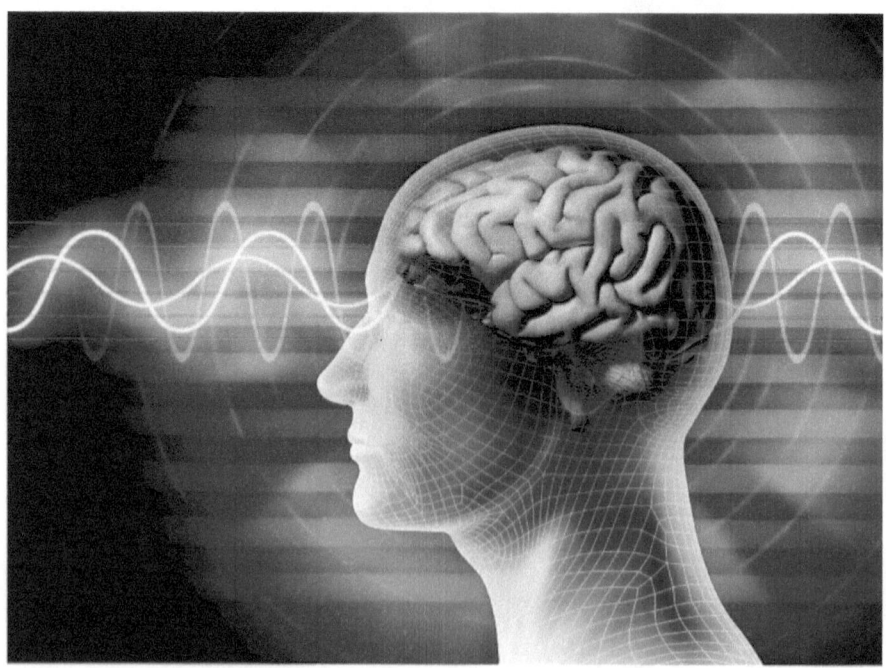

FIGURE 57: HUMAN BRAINWAVES

It's well known that the brain is an electrochemical organ. A functioning brain can generate as much as 10 watts of electrical power. This limited electrical power occurs in precise ways. The electrical activity is displayed in brainwaves, and there are four distinct categories, ranging from the most to least active [19].

Research has shown that one brainwave state may be predominant depending on the time. An activity can result in a mix of all three brain states at all times. So, while somebody is in an aroused state and exhibiting beta brainwaves, there is also a component of alpha, theta, and delta, even if they are only present at a trace level.

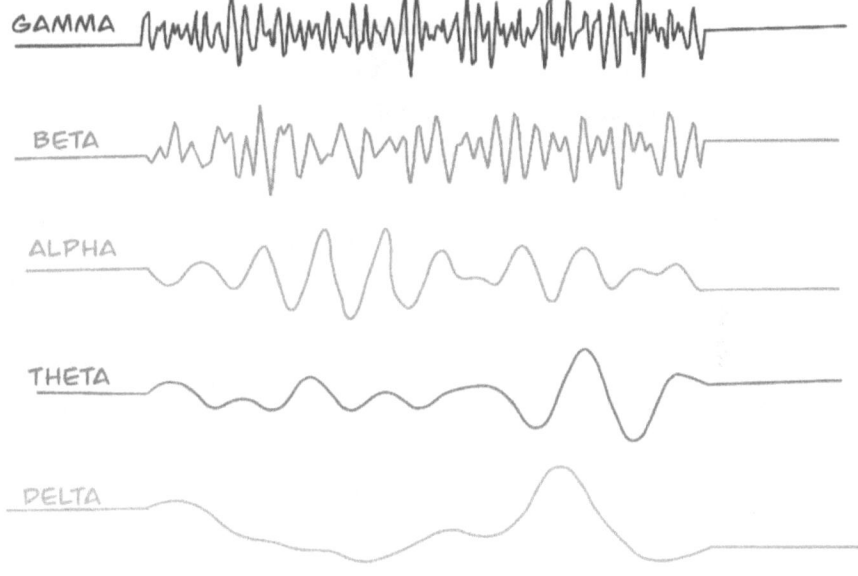

FIGURE 58: DIFFERENT TYPES OF BRAIN WAVES [19]

GAMMA WAVES

It was recently discovered that Zen Buddhist monks show an extraordinary synchronization of their brain waves known as gamma synchrony. This brain state is associated with robust brain function, and these waves have the highest frequency compared to any of the other brainwaves described. Usually, your brain doesn't reach this state unless it's used to intense meditation like the Buddhist monks or intense physical workouts.

BETA WAVES

When the brain is aroused and active, it generates beta waves. A person in active conversation would be in beta. A debater would be in high beta. These waves show an active engagement in work and are relatively low amplitude and are faster than the other four different brainwaves (the frequency ranges from 15 to 40 Hz).

ALPHA WAVES

Alpha waves are the opposite of beta and are a representation of non-arousal and relaxation. These waves ebb slower and have a higher amplitude (the frequency ranges from 9 to 14 Hz). A person relaxed or watching TV (television), or performing meditation would be in an alpha state. Usually, it's essential to maintain this brain state since it's a sign of a healthy mind.

THETA WAVES

Theta waves are usually aroused by the repetitive nature of specific activities such as meditation or listening to a boring lecture. Usually, a person is in this brain state when they are drowsy or about to fall asleep. Theta waves have an even slower frequency and a greater amplitude than an alpha state (4 to 8 Hz). Individuals commonly get a lot of good ideas during this time due to the relaxation of the brain. During an alpha and theta state, the mind is slowed, allowing for a greater flow of ideas, and this commonly occurs in the shower or while brushing hair. It's usually a state when tasks become so automatic that your mind mentally disengages and reaches a free flow status without censorship. This state of mind is typically very positive and a common place that people wish to achieve during meditation or hypnosis.

DELTA WAVES

Delta waves only occur during deep, dreamless sleep, when your brain can reach the lowest frequency possible and the greatest amplitude (frequency 1.5 to 4 Hz). It's important to note that your mind never reaches zero during this time; that would mean that it's dead.

It's known that humans dream in 90-minute cycles. Usually, when delta waves increase to theta, that's when active dreaming occurs and becomes more experiential. This is known as REM sleep or rapid eye movement, which is a characteristic of active dreaming. When you awake in the morning, it's possible to stay in the theta state for an extended period of time, which allows for a large amount of deep relaxation resulting in extremely productive and creative mental activity.

SUPERSONIC SHOCK WAVES

One commonly hears about supersonic shock waves when addressing superheroes or science fiction; interestingly, this is a real-world phenomenon that occurs when an aircraft travels faster than the speed of sound. A supersonic shock wave is a strong pressure wave in an elastic medium such as air or water caused by explosions, lightning, or other phenomena that create violent pressure changes.

During a shock wave, a region of sudden and violent change in stress, density, and temperature results in these waves propagating differently from ordinary ones. Particularly, shock waves travel faster than sound, and their speed increases as the amplitude is raised. Yet, the intensity decreases as the speed increases. The strength of this shock wave also decreases faster because some of the energy is expended into heat.

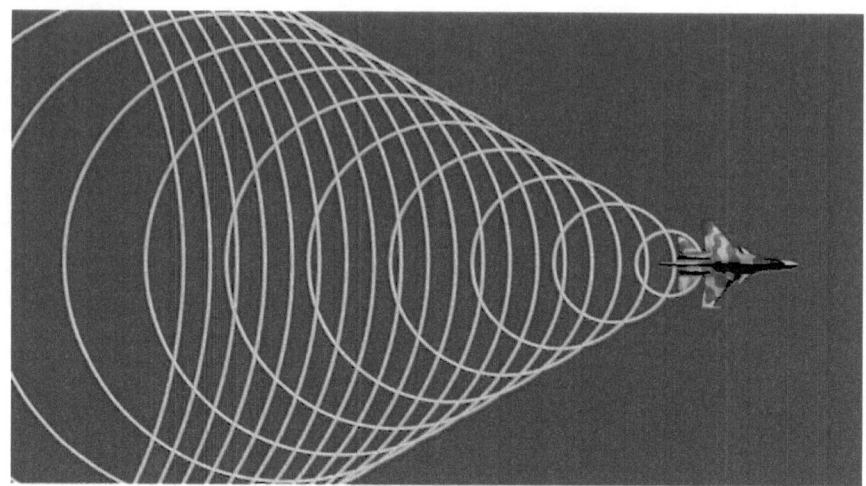

FIGURE 59: SUPERSONIC SHOCK WAVES OF A FIGHTER AIRCRAFT

When a craft travels quicker than the speed of sound (767 mph or 1235 km/hr.), the vessel moves faster than the waves it creates. The air molecules cannot keep up with this speed and begin to compress, thereby creating a rapid increase in pressure in the front of the craft. The result is a new form of a wave (supersonic wave) forming. Though humans cannot see the formation of these shock waves, we can hear them, such as thunder during a lightning storm, which can also sound like a sonic boom.

SOUND UNDERWATER

Sound travels differently in water than air. Unlike electromagnetic waves, sound waves require some medium to travel, such as air water or, at times, solids. Sound waves travel faster in denser materials because particles in the waves will more easily bump into one another. Therefore, sound waves end up traveling much quicker in water than they do in the air. Additionally, sound can also travel farther distances in water than air; this is because the waves' energy quickly gets lost along the way when traveling through the air. While in water the water particles can carry sound better than air.

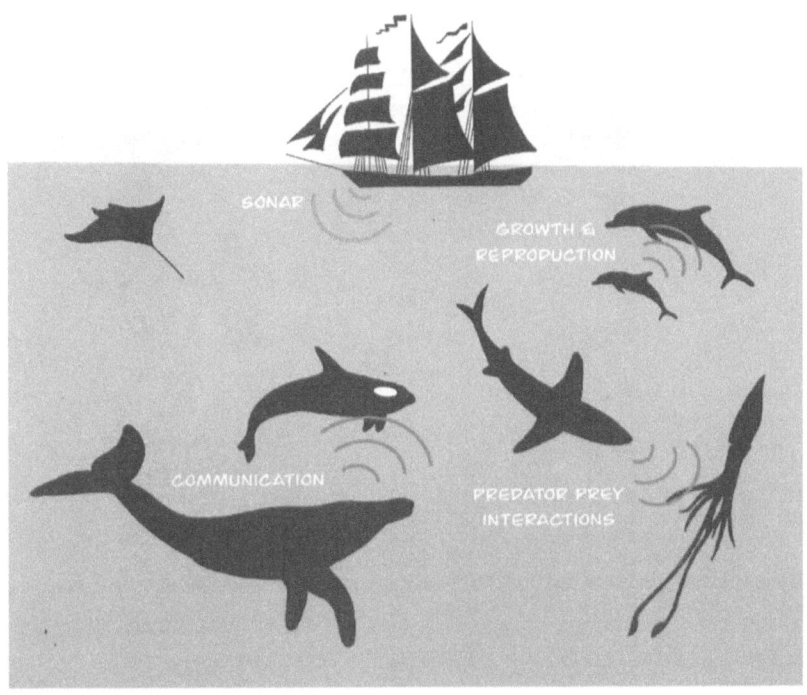

FIGURE 60: SOURCES OF SOUND UNDERWATER

Additionally, sounds also seem louder in water than they do in the air; this is due to the intensity increasing because of sound waves' ability to propagate better in a substance such as water. Sonar is a technological advancement that is also incredibly useful and only possible due to the unusual nature of sound waves in water. It is incredibly challenging to navigate underwater; it's for this reason that ships and animals use sonar to determine their surroundings. Sonar makes use of an echo. Note, an echo in water is much more potent than an echo in the air due to the increasing prominence of sound waves underwater. So when a ship sends out a sound wave, the wave might hit an object and bounce back. When the wave bounces back, the boat can determine if there is an object near its surroundings that it should avoid. Usually, the reflection of the wave is at a different frequency than the original. Depending on the dip in frequency or the time it takes for a sound wave to travel back to its origin, one can determine through the use of sonar how far an object is and its approximate location. Hence, sonar becomes an essential tool when doing underwater exploration.

GRAVITATIONAL WAVES

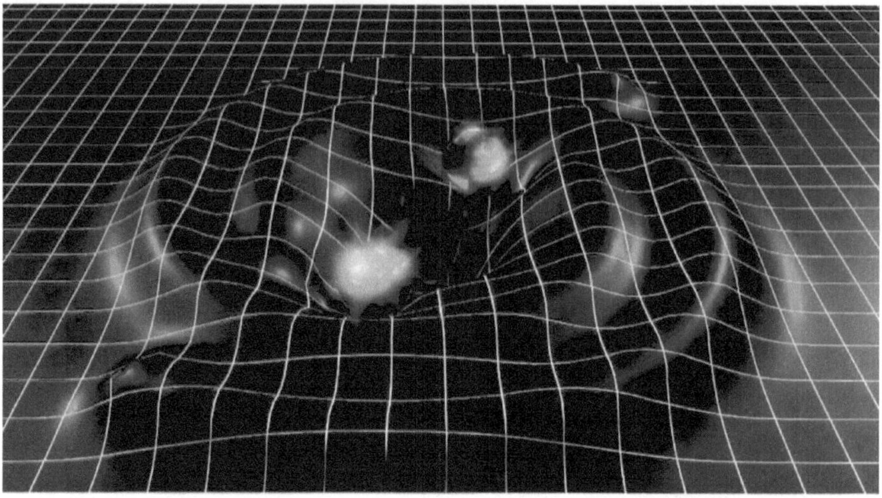

FIGURE 61: SPACE WARPING AND CREATING GRAVITATIONAL WAVES [20]

Einstein first addressed gravitational waves in his general theory of relativity (explained in the later section). Einstein's mathematics shows that cosmic entities would disrupt space-time so that "waves" would propagate around the source. These cosmic waves would travel like ripples in a pond and provide important information to scientists about the logistics of gravity. The most eccentric processes cause such waves in the Universe, such as the collision of black holes or supernova explosions.

Depending on how drastic the event, the stronger the gravitational waves are. But, it is generally known that incredibly massive objects propagate waves of gravity throughout space-time. Though Einstein predicted this in 1916, the proof of their existence was found in 1974. During that time, two astronomers using a radio observatory discovered a binary pulsar, like the one predicted by Einstein all those years ago. After eight years of constant observation, they determined that these waves were coming from two stars moving closer and closer to one another, and therefore were emitting gravitational waves.

Over time, many astronomers and scientists have continued to study pulsar radio-emissions and found similar effects. This research was furthered in 2015, when the LIGO laboratory physically sensed the propagation of space-time caused by gravitational waves, resulting from the collision of two black holes 1.3 billion light-years away. These results were the definitive proof that there were ripples in space-time as predicted by Einstein.

HEAT AND THERMODYNAMICS

The Laws of Thermodynamics are considered the building blocks of how energy exchange works in our Universe today. These laws are assumed to be valid by the majority of physicists. However, due to large amounts of phenomena discovered in the modern era (zero-point energy), these laws are no longer considered as airtight as initially hoped. Instead, these laws cause profound amounts of strife amongst the scientific community, some believing that they are still valid, and others having second thoughts.

TEMPERATURE

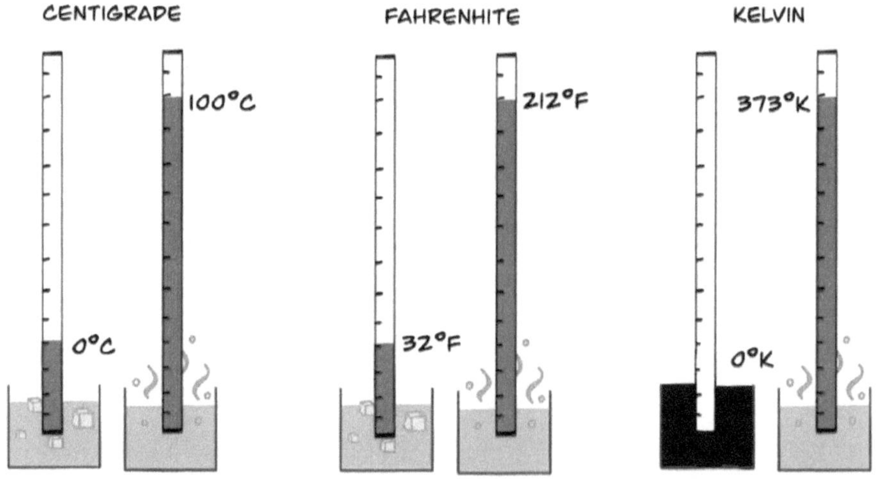

FIGURE 62: DIFFERENT WAYS OF TEMPERATURE MEASUREMENT

Temperature is commonly defined as the measurement of the average kinetic energy of particles within an object. So, as the movement of these particles increases, temperature also increases. A standard piece of confusion is the difference between temperature and thermal energy. Temperature is different from thermal energy. As stated previously, temperature describes the average thermal energy in an object. Thermal energy, on the other hand, represents the total kinetic energy within a substance. Like thermal energy, heat is also different from temperature. Heat focuses on the interaction between objects in a system; temperature is simply measuring the system's internal energy, not its interaction with other systems.

There are quite a few different ways to measure temperature. The three major ones are Kelvin, Celsius, and Fahrenheit. The Celsius scale is based on water temperature, where 0°C represents the temperature at which water freezes and 100°C is where water boils. Kelvin is similar to the Celsius scale; but, it is just shifted by 273.15. This method ensures that there are no negative numbers, which becomes very useful for calculations. The value of 0°K represents absolute zero, which is the temperature at which particles cease to move. Scientists have been trying to reach this temperature; unfortunately, it has still not been achieved. Fahrenheit is one of the least used temperature scales for scientific purposes; it's based on the human body's temperature, usually around 98°F.

The temperature of the triple point of water is taken as 273.16°K or 0.01°C to calibrate thermometers. Triple point water is the temperature and pressure at which water can co-exist in solid, liquid, and gaseous states.

ZEROTH LAW OF THERMODYNAMICS

A lesser-known law of thermodynamics generally referred to as the Zeroth Law of Thermodynamics states that if two systems are in thermal equilibrium, and a third system is introduced, all three will reach thermal equilibrium. Thermal equilibrium is when all three bodies reach the same temperature or an equilibrium value of temperature. An example that we experience daily is putting ice in a glass of water at room temperature. After some time, the ice melts, and the entire glass of water will reach an equilibrium temperature. This principle sounds like common sense or a shared experience, but it is essential to note this phenomenon happens across all bodies and the entire universe.

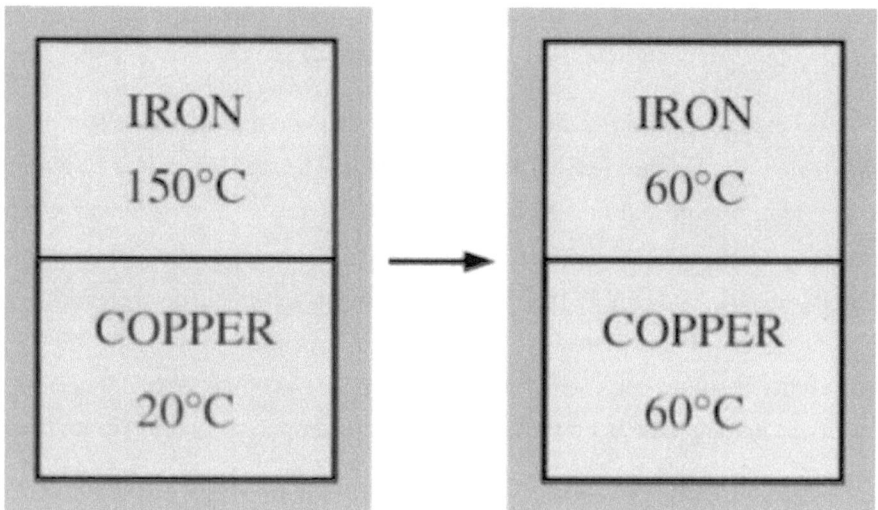

FIGURE 63: MOVING TO EQUILIBRIUM TEMPERATURE

Temperature change occurs when mixing hot coffee with cold milk. When the two liquids mix, an equilibrium temperature is reached, in which the coffee is no longer as hot as it used to be, and the milk is warmer than before. This common phenomenon is witnessed early in the morning, and the Zeroth Law of Thermodynamics defines why this occurs.

FIRST LAW OF THERMODYNAMICS

The First Law of Thermodynamics (Law of Conservation of Energy) states that energy cannot be created or destroyed for isolated systems. It is imperative to draw a system boundary to validate this law. The energy calculations need to be performed inside the system boundary, and there should not be any energy transfer across this boundary. This law is sometimes considered a special case of conservation of energy mainly used for heat and thermodynamic phenomena. This is a universal law, and it is valid in all situations and with all types of energy conversions. In any phenomenon or experiment, the total input energy has to be equal to the total sum of the output energy.

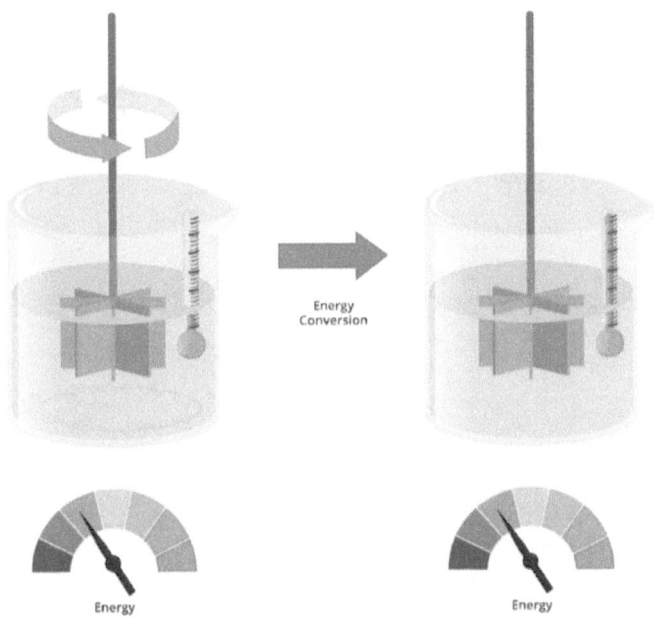

FIGURE 64: CONVERSION OF MECHANICAL ENERGY TO HEAT

1. According to this law, no machine would continue running forever. Therefore, perpetual motion (e.g., a wheel rotating by itself without the application of energy or force) does not exist.
2. In a motor, electrical energy is transformed into mechanical energy, sound, and heat. Due to this, the input and output energy are equivalent. Mechanical rotational energy is used to move the car; the other outputs such as sound and heat are not as useful.

HEAT

As explained earlier, temperature can also be considered a physical property for all matter. It represents the number of oscillations or the movement of atoms and molecules in a substance. When molecular oscillations decrease, the temperature decreases (the movement of molecules slows down), and when the oscillations increase, the temperature increases (the movement of molecules increases). Additionally, when two bodies of different temperatures are brought near one another, the hotter body gets colder, and the cooler body gets hotter until they're both at thermal equilibrium (Zeroth law of thermodynamics). Heat is a form of energy transferred between bodies until they both have the same temperature. As mentioned earlier, in the laws of thermodynamics, heat generally moves from hot to cold. The temperature at which the molecule oscillations come to a complete halt is absolute zero (-273 Celsius), which is almost impossible to achieve.

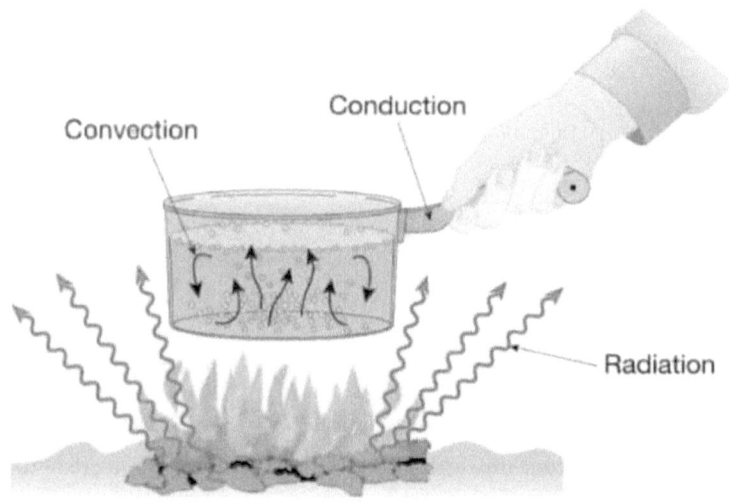

FIGURE 65: DIFFERENT FORMS OF HEAT TRANSFER

There are mainly three ways heat is transmitted: conduction, convection, and radiation. All the heat transfer phenomena always follow the zeroth law of thermodynamics, where heat transfers from a cold body to a hot body.

Conduction (or thermal conduction) happens because of molecular and atomic movement, resulting in solids, liquids, and gases. Conduction is a common way the heat transfer occurs in solids. This form of heat transfer happens because the amplitude of vibration of the atoms increases as they are heated (e.g., a metal rod is heated). These atoms transfer the vibrational energy to the nearby atoms and this, in turn, conducts heat.

Convection arises because of fluids and is the main form of heat transfer in liquids and gases. We experience convection every day. When you place your hand near a burning candle, you feel the hot air nearby. This transfer of heat is because of convection. In our atmosphere, convection plays a significant role in changing global temperatures and climate patterns on Earth. Also, oceans transfer a tremendous amount of energy through convection.

Radiation happens because of electromagnetic waves. Note heat is also a form of electromagnetic radiation. Radiation can happen everywhere and even in a vacuum. The Sun's energy reaches Earth through radiation. Mammals like rattlesnakes use this phenomenon of radiation to attack their prey. These rattlesnakes have thermal sensors to detect the heat radiation from their game to attack and inject their venom. Also, when radiation falls on a portion of a body, it is absorbed, part of it is reflected, and the remaining portion is retransmitted.

It is found that shining bodies reflect maximum, and the black bodies are nice absorbers of heat. Surprisingly, bodies that are transparent to light are also good transmitters of heat.

THERMAL EXPANSION

Thermal expansion is the general increase in the size of a material as its temperature grows. Linear expansion is when the length changes in comparison to the original amount. There can also be a real expansion, where the area changes during the process of enlargement. And finally, there is also volumetric expansion, where the volume of the substance changes.

For most materials, the fractional shape changes due to heat are directly proportional to the change in temperature over a small temperature range. They also have the same sign, which means that materials expand when heated and contract when cooled. The ratio of the change in the size of a material to its temperature change is known as the coefficient of thermal expansion.

EXPANSION OF SOLIDS

When a solid is heated, the atoms that make up the substance vibrate faster around their fixed points. Hence, the thermal expansion of solids is relatively small, and these solids also tend to retain their general shape. A typical example is metal railway tracks, which have small gaps so that when they are heated by the Sun and the rails expand, they don't ruin the railroad.

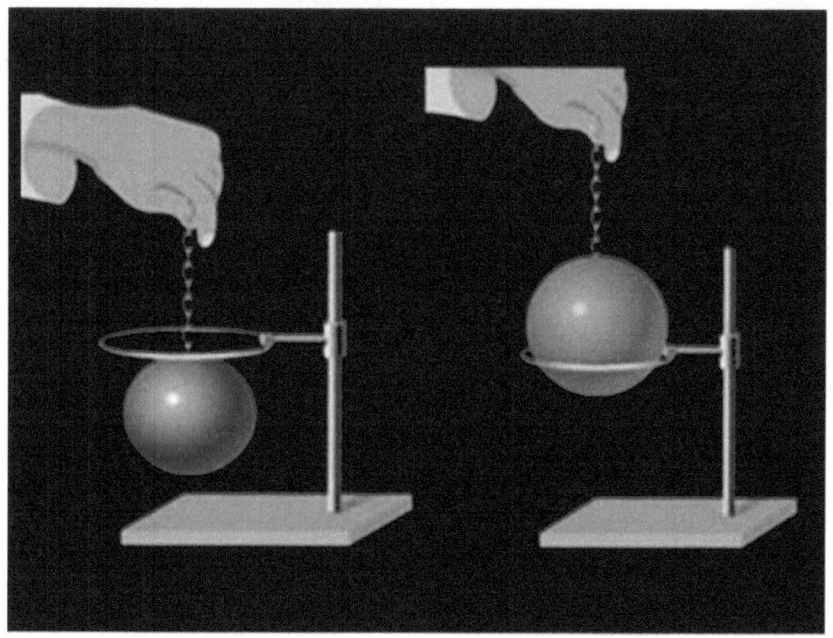

FIGURE 66: SOLIDS EXPAND WHEN HEATED

EXPANSION OF LIQUIDS

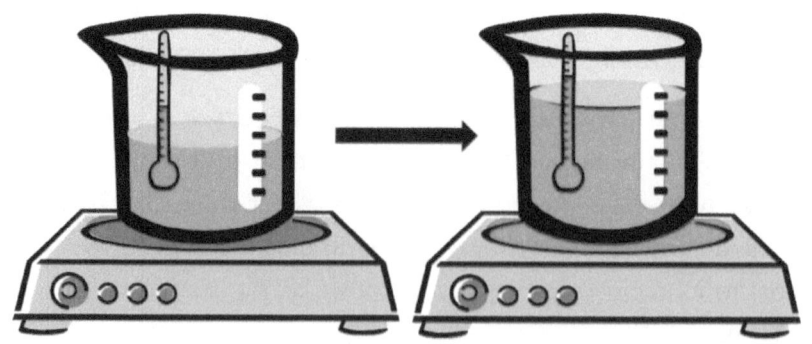

FIGURE 67: LIQUIDS EXPAND WHEN HEATED

Liquids expand for the same reason as solids; however, they expand more than solids because the bonds between liquid molecules are less tight.

Therefore, when liquids are heated, kinetic energy is added to the molecules, causing them to move faster on average and move further apart and collide. Liquid thermal expansion is the same reason why liquid-in-glass thermometers work. As the thermometer heats up, the liquid expands and rises in the glass tube.

ANOMALOUS EXPANSION OF WATER

Water behaves strangely when cooled from room temperature to 0°C. Water contracts till it reaches 4°C, and from 4°C to 0°C, it expands in volume. This behavior of water is called the anomalous expansion of water, and hence water retains the highest density at 4°C as the volume is lowest for a given mass of water.

This behavior of water makes the lakes and oceans freeze from top to bottom rather than bottom to top. As the temperatures go down, the lake's top layers temperature keep decreasing. The warm water from the bottom with lesser density reaches the top until the entire lake's temperature starts becoming 4°C. As the water temperature decreases further, the lake's top layers cool down to 4°C, and gradually the density reaches the maximum value. Soon, the entire lake temperature becomes 4°C. Upon further cooling from 4°C to 0°C, the water density decreases as the volume increases. Hence, the top layers of water do not go down as the density is lower than the water below, and in turn, only the top layers freeze first at 0°C. When the top layers freeze first, they act as an insulator of heat to the bottom layers, delaying the freezing of bottom layers of water. This property of water saves marine life as the bottom layers don't freeze during winters.

HEAT RADIATION

Almost all objects, including the Earth, have some ability to absorb heat. The majority of this heat comes from the Sun in the form of thermal and light energy. Similarly, the Earth's atmosphere also absorbs heat from the Sun, and this makes it an incredibly important aspect of our life.

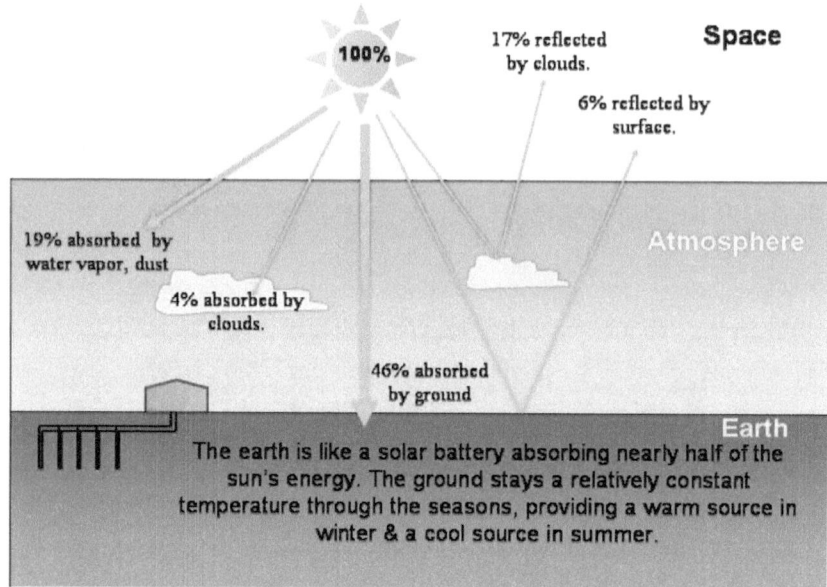

FIGURE 68: REFLECTED AND ABSORBED ENERGY OF SUN

Like how objects on Earth absorb light and heat, objects also absorb heat from other stars and celestial bodies within space. Hence, heat and thermal radiation is a constant form of energy which is always consumed by cosmic entities throughout the universe.

STEFAN-BOLTZMANN LAW

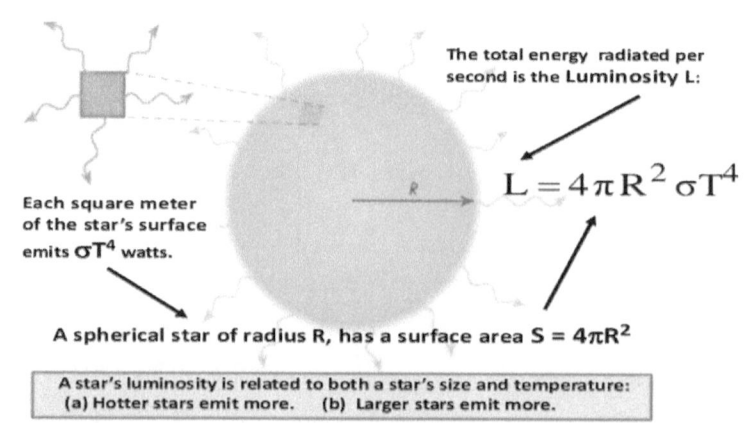

FIGURE 69: CALCULATION OF RADIATED ENERGY FROM A STAR

Formulated in 1879 through experimentation by Austrian physicist Josef Stefan and derived mathematically in 1884 by physicist Ludwig Boltzmann from thermodynamic principles, the law describes the radiated heat power emitted from a surface. This law makes it possible to determine the radiated energy from a star or any object. It is commonly used to determine the heat of a star and the total energy radiating into space, making this law unique to astronomers. This law states that the amount of radiation from a body mainly depends on the temperature and surface area. Note this law can be applied to any surface, and the surface need not be a sphere.

WIEN'S DISPLACEMENT LAW

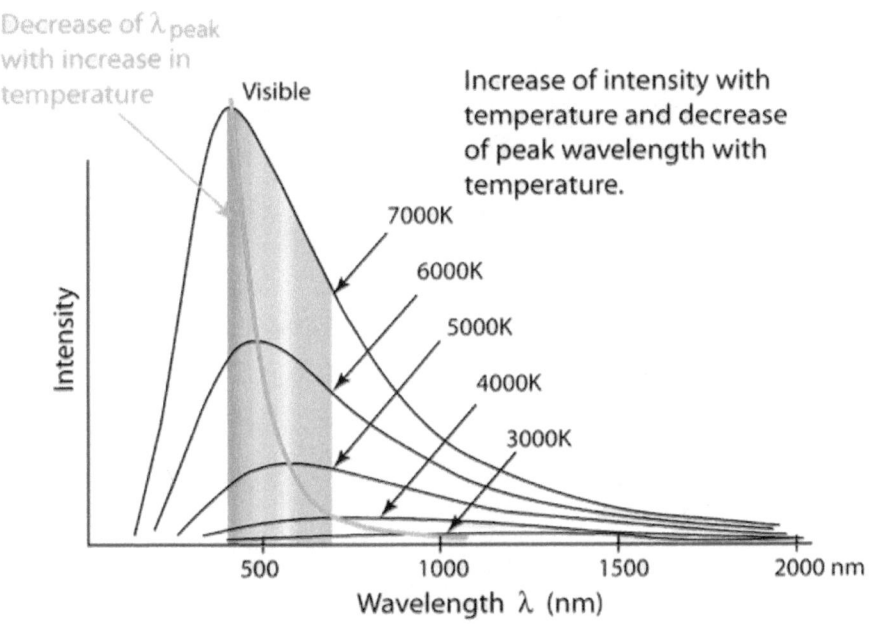

FIGURE 70: WEIN'S WAVELENGTH DISTRIBUTION & TEMPERATURE
[24]

Wein's displacement law describes the relationship between a blackbody (another term for a perfect substance that absorbs all frequencies of white light) and the wavelength at which it emits the most light. Wein used an oven with a small hole as a representation for an ideal blackbody. This system ensures that nearly all radiation is absorbed since any radiation entering the hole will be scattered and reflected by the oven walls. So, all of the light was absorbed, similar to that of an ideal blackbody.

Wien found that as the temperature is increased, the blackbody emits shorter wavelengths. This phenomenon is also commonly witnessed. Warm objects emit infrared radiation and can be observed with a dull red glow. But, the hotter a substance gets, the brighter the light it radiates. Based on this law, it has been found that:

1) Black bodies radiate all the wavelengths when heated.
2) All black bodies emit the same spectrum of radiation at a given temperature.
3) As the temperatures increase, the radiation intensity at every wavelength also increases.
4) The wavelength of the radiation with maximum power is inversely proportional to the temperature.

NEWTON'S LAW OF COOLING

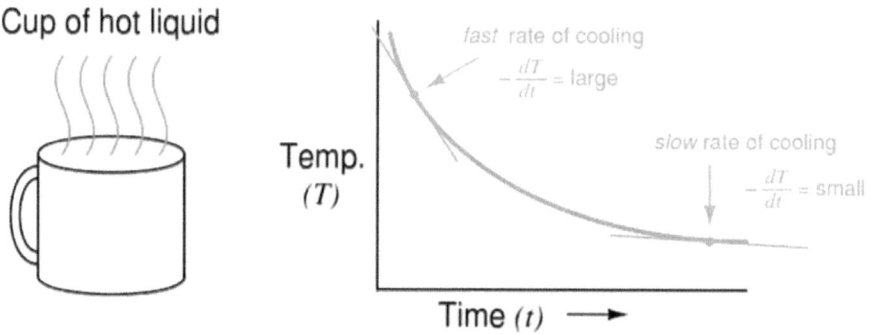

FIGURE 71: RATE OF COOLING DECREASES WITH LOWER TEMPERATURE

Newton's Law of Cooling defines the rate at which an exposed substance changes temperature through radiation. According to this law, the temperature change is proportional to the difference between the object's temperature and surroundings.

This idea seems like common sense. If you have a hot coffee cup and put it in a refrigerator, it will cool faster than if you place the same cup on a table. Since the environment within the refrigerator is much colder than the table, and the cup is boiling, hence the cup will cool faster and proportionally to its environment. Now, if you were to put a cold cup of iced tea within the fridge, the temperature change would be much smaller than with a hot coffee cup, the reason being that the temperature of the iced tea and fridge is quite similar.

ENTROPY

The concepts of energy and entropy become incredibly important to understand when discussing thermodynamics. Entropy can be a confusing subject; but, in the simplest of terms, it's the number of ways that the energy distribution changes throughout a system, though the total amount of energy in a system remains constant. Note the first law of thermodynamics only states that the total energy is conserved in an isolated system. Still, it does not talk about how energy is distributed within the system.

LOW ENTROPY

A IMPROBABLE ARRANGEMENT OF MARSHMALLOWS

HIGH ENTROPY

A PROBABLE ARRANGEMENT OF MARSHMALLOWS

FIGURE 72: ENTROPY WITH MARSHMALLOWS IN A COFFEE CUP

A system consists of many microstates or microsystems. The microstates will have some measurable properties like temperature, pressure, volume, etc., which are the byproducts of the microstates' internal energy. Suppose all these microstates have an equal amount of energy (which is highly unlikely), then the entropy of the system will be very low.

To further explain, everybody or any system has a certain amount of energy. For example, a system such as a glass of water has a certain amount of energy, which causes specific physical properties such as temperature, pressure, and volume. Every water molecule in the container contains some amount of energy (unequal amounts), where their sum is the total energy contained in the glass of water. The distribution of energy among these water molecules continuously changes as molecules and atoms exchange tiny amounts of energy; but, the total energy in the glass of water remains constant. Entropy is the number of permutations of the energy states of the system. For example, there is a higher energy transfer in liquid water than in an ice cube since the molecules can move more freely. Therefore, the liquid water has higher amounts of entropy than the ice cube. This basic understanding of entropy is critical for the laws of thermodynamics.

SECOND LAW OF THERMODYNAMICS

The Second Law of Thermodynamics (Law of Entropy) states that the entropy of an isolated system always increases with time. This law is a property of an isolated system in the universe to exchange energy and reach an average equilibrium temperature by increasing the overall entropy. Therefore, the entropy of the universe is always increasing. Hence, the energy distribution of the universe changes accordingly to ensure that the entropy always increases.

A reversible process is one in which the entropy of the system goes back to the previous state. The probability of a reversible process is extremely low, and in the universe, most processes are irreversible where the entropy always increases. Note a macrosystem consists of many microsystems where most processes are irreversible, and hence the entropy always increases in a macrosystem. Also, note entropy is a function of the state of the system or the energy distribution of its microstates.

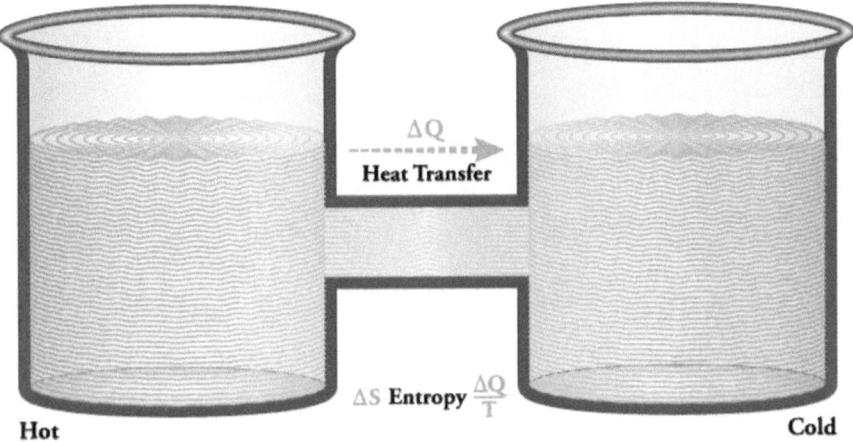

FIGURE 73: TRANSFER OF HEAT FROM A HOT BODY TO A COLD BODY

1. For example, think of an isolated system consisting of a glass of hot water and a glass of ice cubes. When these two glasses are near one another, the ice cubes melt, and the hot water temperature decreases, thereby increasing the entropy of the whole system since there is an energy exchange between the two glasses. There is a small probability of a transfer of energy happening in which the ice cubes get colder and the hot water gets hotter. Yet, upon experimentation, this type of transfer doesn't occur because its probability is too small. The most likely occurrence is that the ice cubes melt and the hot water gets colder in turn, increasing the entropy.

2. When an ice cube is put into a cup of water, the water gives off heat (the most probable energy exchange), which results in the ice cube melting over time. The ice cube absorbs the thermal energy given off by the water, so the entropy increases. Since the ice cube is always at a lower temperature than the water, the entropy is continuously increasing within the system.

THIRD LAW OF THERMODYNAMICS

The entropy of a pure crystalline substance at absolute zero is 0.

$$S = k\, lnW = k\, ln1 = 0$$

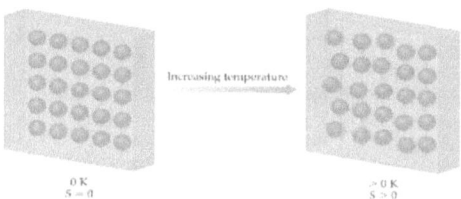

FIGURE 74: PURE CRYSTALS HAVE ZERO ENTROPY AT ABSOLUTE ZERO TEMPERATURE [21]

Entropy approaches a constant value as temperature approaches absolute zero. Therefore, the system will have the least amount of energy at absolute zero. In the picture, if you don't understand the mathematical equation, it is okay; just focus on the written content.

The important thing to understand is that as an object's temperature cools, the motion amongst the molecules and energy transfer decreases, resulting in a direct reduction in entropy. Absolute zero is when there's no motion in the molecules that make up a substance (scientists have gotten close but haven't reached this temperature yet).

KINETIC THEORY OF GASES

The kinetic theory of gases is focused on the molecular movement and description of gas. This theory is based on three key assumptions:

(1) Gas consists of a high number of identical molecules moving in random directions. These molecules are separated by large distances in comparison to their small size.

(2) These molecules undergo perfectly elastic collisions, which means that no energy is lost when they interact with each other and the container walls.

(3) When these molecules collide with one another, the kinetic energy transfer between them is only in the form of heat.

This model has been known to describe a perfect gas, which can be considered a reasonable approximation for an actual gas. However, this theory fails when accounting for the behavior of gases at high densities.

According to this theory, the pressure on the walls of any container can be quantitatively associated with the random collisions of the gas molecules inside the container with the container's walls and themselves. Hence, the more that these molecules collide, the more pressure there is inside this container. The two factors that determine how much collision happens within gas molecules are the temperature and density.

If there is a higher temperature of the gas in the container, then the molecules move faster, which results in them bumping into each other more often, thereby increasing pressure. The same is valid for increasing the density. If there are a higher number of gas molecules inside the container, there will be more collisions, resulting in higher pressure. Inherently, this kinetic theory of gases, temperature, pressure, and density are closely related to one another and affect each other through a series of relationships.

FIGURE 75: BROWNIAN MOTION OF GAS MOLECULES [25]

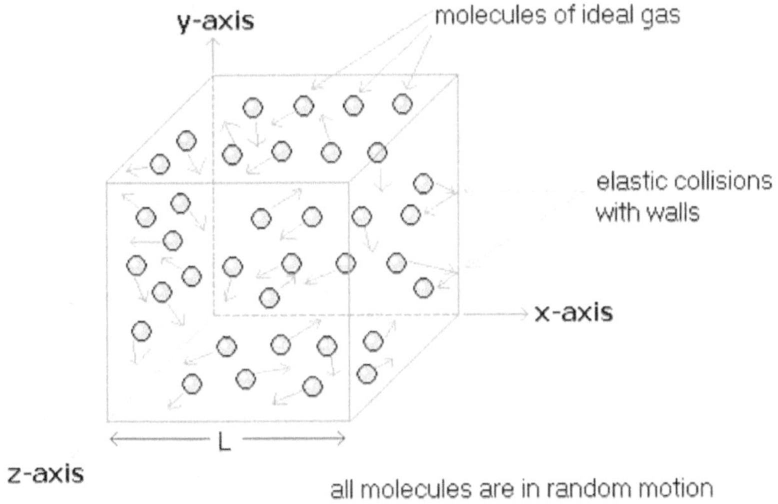

AVOGADRO'S NUMBER

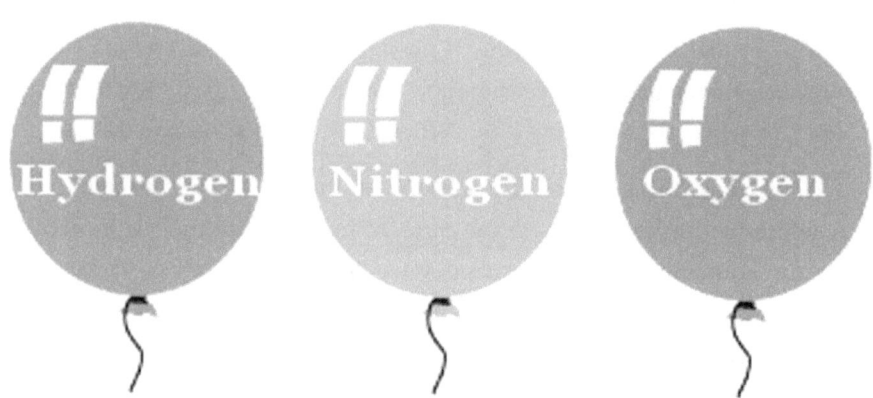

The balloons all have the same volume. This means they all contain the same number of molecules.

FIGURE 76: GASES OF EQUAL VOLUME CONTAIN EQUAL NUMBER OF MOLECULES

In 1811, Amedeo Avogadro had the idea that if you had equal volumes at the same temperature and pressure, they would contain an equal number of particles. Unfortunately, during his time, most scientists viewed Avogadro's work as being purely hypothetical since there was no exact way to determine the difference between atoms and molecules.

Later, by the late 1860s, Avogadro was proven correct (unfortunately, he died in 1856). Avogadro's number describes the sheer amount of particles in a space at $0°C$ and the pressure of one atmosphere. Imagine a balloon filled with gas. There are precisely six hundred and two sextillion gas particles (6.02×10^{23}). The same can be shown through water molecules. If one were to pour 18.01 mL or a mole of water into a glass, you have the same amount of water molecules (6.02×10^{23}). Since Avogadro was the first to coin the idea, this number was named after him, known as Avogadro's number.

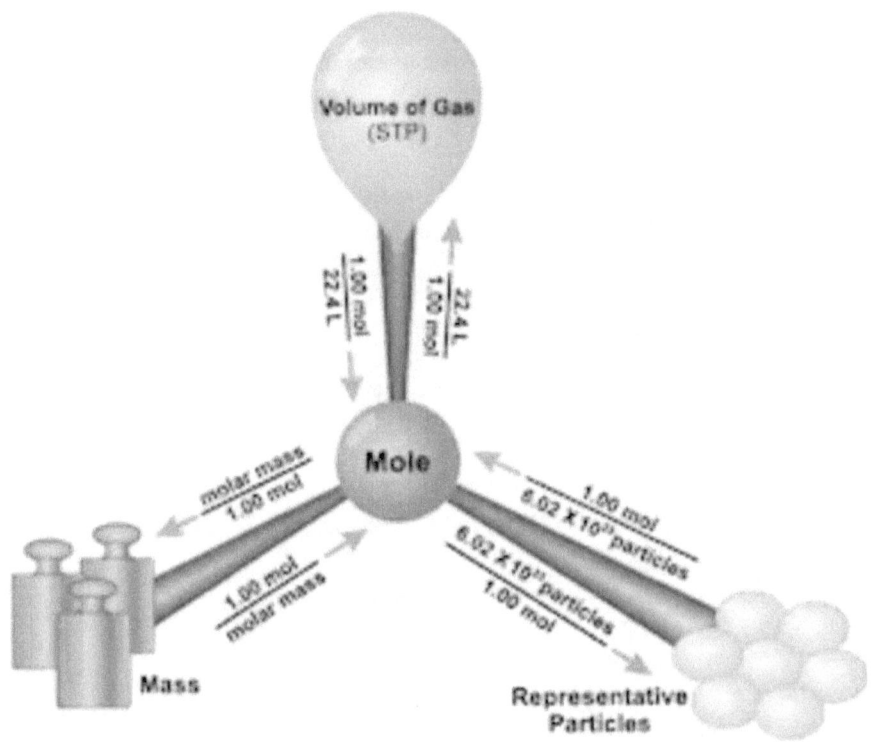

FIGURE 77: AVAGADRO'S NUMBER AND MOLAR MASS

Avogadro's number is also known as the mole in chemistry. Scientists use the mole to determine the number of particles in a volume with magnitudes of 602 sextillion. These are also known as molar quantities. Since atoms and molecules are so small, they have to be bundled into groups called moles. One would use the mole the same way one would use pounds or grams. It's merely another way to measure incredibly large quantities.

EQUIPARTITION OF ENERGY

The equipartition of energy, also known as the law of equipartition, is a law in thermodynamics that states that an equal amount of energy will be associated with each independent energy state in a system in thermal equilibrium. I know that sounds complicated; but, it's quite simple.

Let's say that there is an atom that can absorb energy in three different ways. Perhaps it's near three separate heat sources. The unique ways an object can acquire energy are known as the degrees of freedom, thereby giving this specific atom three degrees of freedom. The law is then used to calculate the average amount of kinetic energy in every molecule in the system, based on the degrees of freedom in that one atom.

Overall, there are various formulas created by Physicists Maxwell and Boltzmann to describe the energy of atoms in various systems. But, they all follow the same basic principle outlined above.

ELECTRICITY AND MAGNETISM

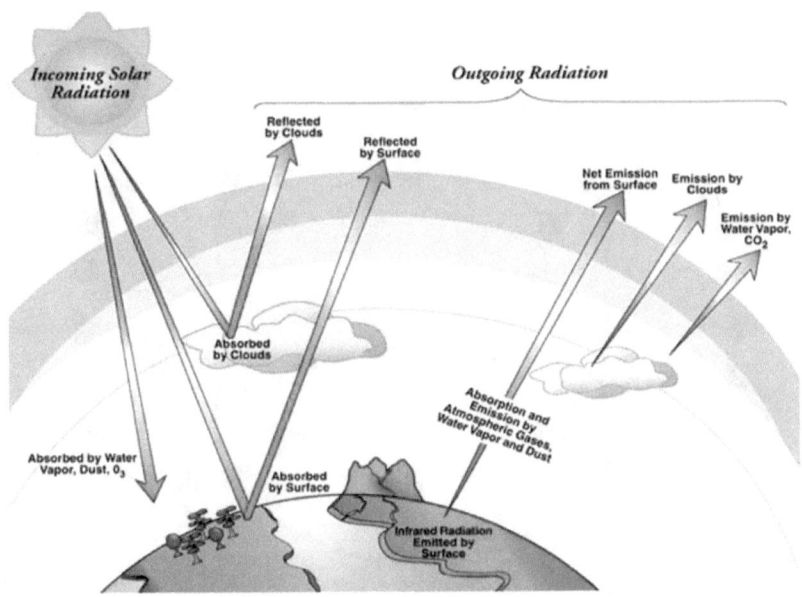

FIGURE 78: INCOMING AND OUTGOING RADIATION FROM SUN

There are numerous different forms of energy. Some of the most common forms are heat, sound, vibrations, light, electromagnetic, chemical, and nuclear. This chapter will analyze the different forms of energy, fields and the physical laws behind them [22].

FIELDS AND FIELD LINES

Field and field lines are a property of the universe which are not entirely understood. Nonetheless, a field is a region of influence due to some force or charge, and it consists of invisible field lines that propagate in the direction of the field's influence. The density of these field lines, also referred to as field flux, is used to measure its strength. These lines never intersect; they bend appropriately depending on the strength and direction of the field. In nature, the most commonly observed fields are gravitational, electric, and magnetic. A typical example is a magnet attracting an iron object; this attraction occurs due to the magnetic field being emitted by the magnet.

Though some physicists believe that the entirety of the universe is made based only on these three fields, others believe there may be other fields that have not yet been discovered.

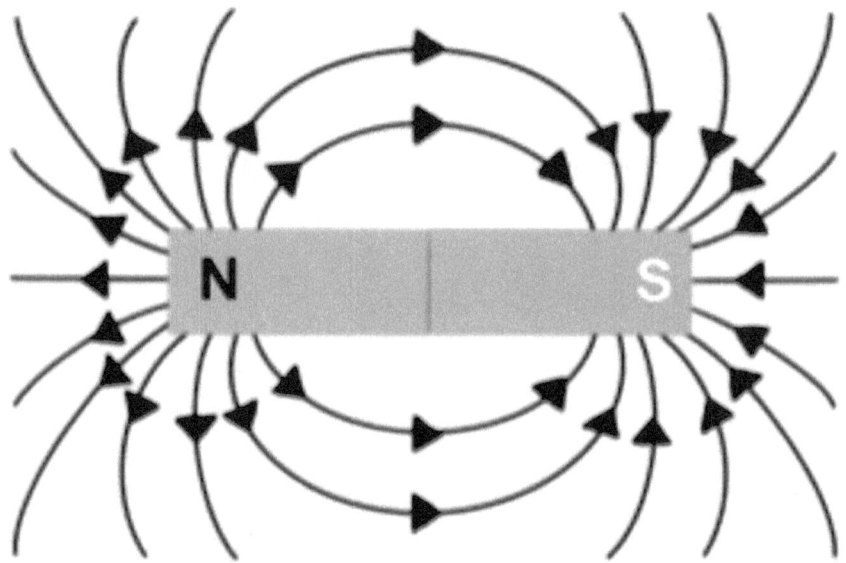

FIGURE 79: POLES AND MAGNETIC FIELD LINES OF A BAR MAGNET

Along with these fields' existence, there is also a general understanding of how to manipulate them. For example, it is known that the same type of electrical charges repel each other, and opposites attract. In terms of magnetic fields, the same poles repel, and the opposite poles attract. This commonality between electricity and magnetism has been seen quite consistently in nature. There is something known as the electromagnetic force, which leads to several phenomena associated with electricity and magnetism. For example, a moving electric charge generates a magnetic field and an electric field (this will be elaborated further later in this book).

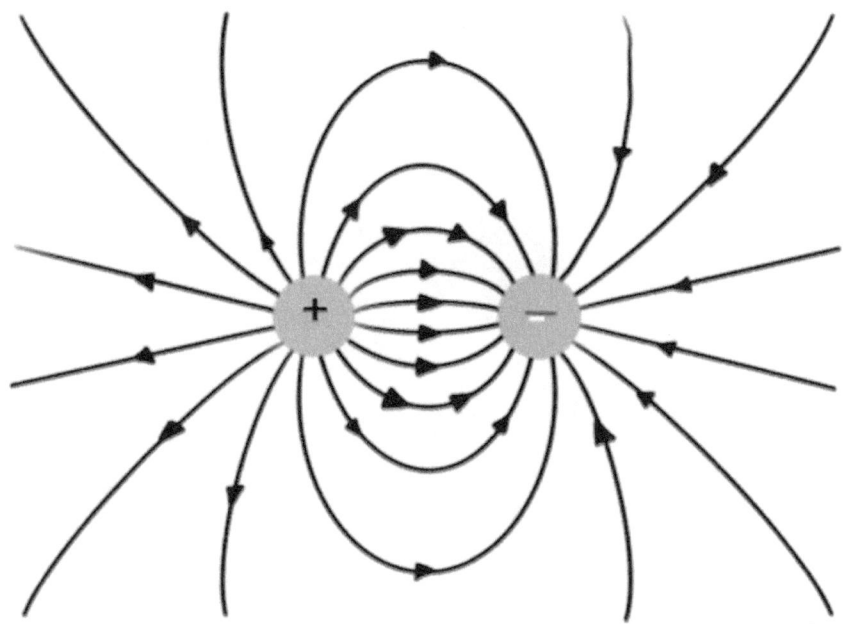

FIGURE 80: ELECTRIC CHARGES AND ELECTRIC FIELD LINES

Electric and magnetic fields can be attractive or repulsive, whereas gravitational fields are only attractive. These phenomena result in a widespread question regarding fields (other than what causes their existence), how an object in a field understands the field's presence, or the changing strength of the fields—going back to the example of a magnetic field and an iron object. It is still unknown how the iron object recognizes the generated magnetic field. The same goes for objects in Earth's gravitational field. How does an object know that it is being affected by this field? There are several theories to explain this phenomenon, which we will elaborate on further later. In short, we cannot see these fields or field lines, but we can perceive their presence through experiments. We also experience Earth's gravitational field every day. We cannot touch these gravitational field lines, but we share their effect. These field lines are directed inward radially to the center of the Earth.

FIGURE 81: EARTH'S GRAVITATIONAL FIELD LINES

ELECTRICITY

Electricity is a way to describe the flow of electric charge or electrons. Electrons are negatively charged atomic particles. When they exist in an atom, they create a negative charge. Note, upon the removal of electrons, a positive charge is produced in the atoms. Atoms are generally neutral, with an equal amount of positive and negative charges. To further understand electricity, it's essential to be familiar with the following electrical phenomena: electric charge, electric field, electric dipole, electric flux, electric potential, conductors, capacitance, current, resistance, and inductance.

ELECTRIC CHARGE

Mild electrical sparks a common occurrence in everyday life, such as when one touches a doorknob or rubs a balloon against their head. This shock occurs because the electrical charge flows from a positively charged body to a negatively charged body. This is referred to as the transfer of electrons, and, in certain situations, outer electrons in an atom are loosely bonded. This loose bond allows them to be transferred to other atoms easily.

Electric charge is the excess of electrons (e.g., a negative charge) or the loss of electrons (e.g., a positive charge). Therefore, the electric charge can be positive or negative.

The smallest known charged particles would be an electron, proton, or positron (these are all atomic particles, for more information, refer to the chapter detailing particle physics). This electric charge is quantized (we will discuss this in more detail in a later chapter). Additionally, the transfer of electrons in the atmosphere results in Earth's overall negative electric charge of about 500,000 Coulombs spread throughout the surface.

ELECTRIC FIELD

All electrically charged particles create an electric field. An electric field is a region of space where other electrically charged particles are influenced. For example, when two positively charged particles are brought near one another, they repel. Similarly, when two negatively charged particles are brought near each other, they also repel. Only positively and negatively charged particles attract one another; this is where the famous saying, opposites attract, and likes repel comes from. The reason that positive and negative objects attract is because of the electric field lines that are created. In this circumstance, the electric field lines go outward from a positively charged particle into a negatively charged particle resulting in attraction (refer to the image below).

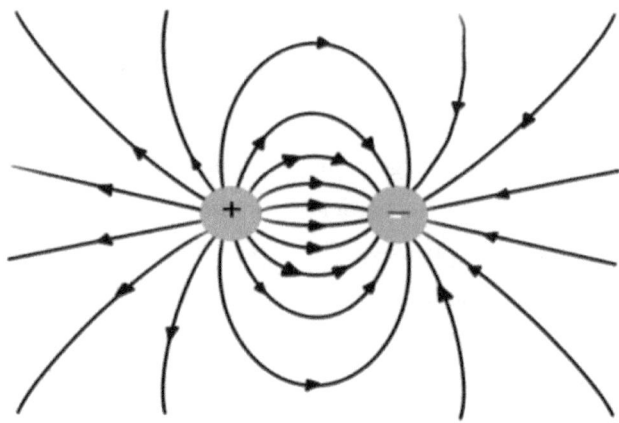

FIGURE 82: POSITIVE AND NEGATIVE CHARGES

ELECTRIC DIPOLE

As stated earlier, any electric charge (positive or negative) generates an electric field. An electric dipole exists if some distance separates a positively charged particle (e.g., a proton) and a negatively charged particle (e.g., an electron). In this situation, the electric field lines emitted from the proton get absorbed by the electron. An electric dipole is used in various electromagnetic applications and is also useful in calculating the potential energy of atomic bonds. Additionally, even an electric dipole carries some potential energy since attracting charges are kept at a distance.

ELECTRIC FLUX

As stated previously, electric fields consist of a large number of electric field lines. The strength of these lines is measured using electric flux. Electric flux is the number of field lines in a given area and is used to measure the electric field strength or the number of electric field lines per unit of area.

ELECTRIC POTENTIAL

In an electric field, a positively charged particle will start moving in the direction of the field. Therefore, a force is acting on the charged particle, and the particle is doing some work. The amount of work done per unit of charge is the electric potential of the field, and the electric potential at different points in an electric field can be varied. The electric potential difference between two points in an electric field is called the potential difference. The potential difference between two areas in effect provides us with a measure of the energy needed to push a unit of electric charge between two points.

We are familiar with the terms voltage in our household through power points (120V in the US) or batteries (e.g., 1.5V). Voltage is nothing but the electric potential difference of two points (e.g., marked + and - in a battery).

Interestingly, our planet Earth has an electric field. This electric field strength changes with altitude, which means that there is a potential difference between various points in the atmosphere. Additionally, this potential difference changes during distinct seasons and also in other regions. This electric field strength is at its peak on the Earth's surface and gradually goes to zero as we reach higher altitudes (10 km from the surface of the Earth, the electric field is negligible).

CONDUCTORS AND INSULATORS

Conductors are materials through which charged particles can move freely (e.g., copper wire, iron, human body, etc.). On the other hand, insulators, or nonconductors, are materials in which charged particles cannot move freely (e.g., plastic, glass, etc.). The properties that determine conductors or insulators are based on the availability of free electrons (or loosely bonded electrons) available in the atomic structure. In metals, due to the availability of loosely connected electrons, they conduct electricity easily. Note that there is still a meager amount of charge that moves in insulators, but it's assumed to be zero for practical reasons.

In summary, in conductors, the charge moves freely, but there is always a small amount of resistance to the movement. As the temperature of the conductor decreases, the resistance also decreases, and conductivity increases. Also, there are certain materials in which when they reach a critically low temperature, the resistance suddenly becomes zero, and they become superconductors (e.g., mercury, lead, etc.)

CURRENT AND RESISTANCE

Electric current is the rate of flow of electric charge with respect to time; this only occurs when the charge is flowing in the form of electrons. A charge is positive, and electrons are negative, so if electrons are flowing in one direction, then the charge or current is flowing in the opposite direction. The flow of current will be different in copper compared to that of glass. This difference in the rate of flow is because of the resistance associated with the material. Different materials have a varied amount of resistance for the flow of current.

OHM'S LAW

The relationship between potential difference (voltage), current, and resistance can be expressed with a simple equation; this is known as Ohm's Law:

Potential Difference (ΔV) = Current (I) x Resistance (R)

According to Ohm's Law, the current traveling through a device is always directly proportional to the electric potential difference applied to the object. For Ohm's Law to be consistent in a machine, there needs to be a linear relationship between current and potential difference. Though this is true in most cases, since most conductors obey Ohm's Law, there are exceptions. For example, modern microelectronics depend entirely on devices that do not obey Ohm's Law.

CAPACITANCE

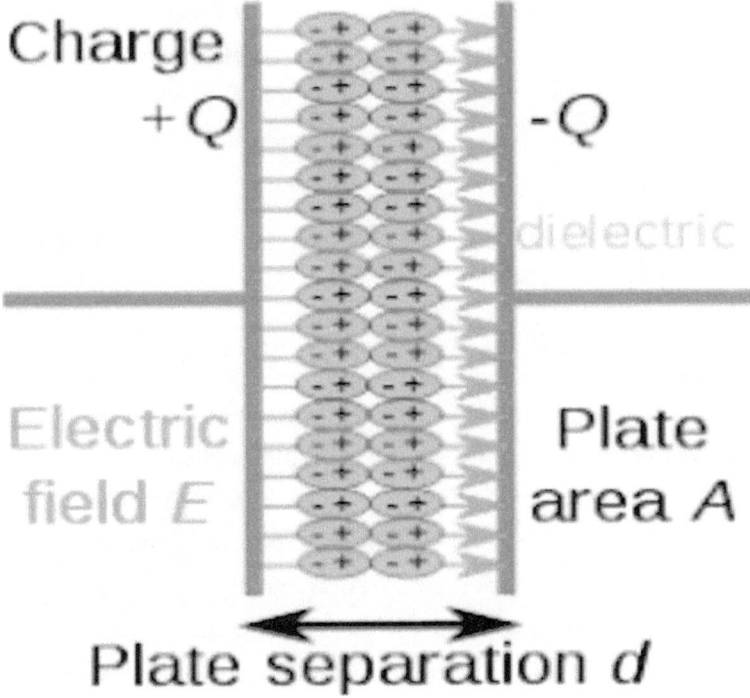

FIGURE 83: CAPACITOR STORING ELECTRIC FIELD [26]

Capacitance is a component or circuit's ability to collect and store energy (usually in the form of electrical charge). Therefore, capacitors are energy-storing devices that consist of two plates of conducting materials (usually thin metals) sandwiched between an insulator. A capacitor collects energy (voltage) as current flows through an electrical circuit, so even when the circuit is switched off, a capacitor retains the energy it has gathered.

Capacitance is mathematically expressed as the ratio of each conductor's electrical charge over the potential difference (voltage) between them. There are several different ways to increase capacitance: a capacitor's plates are positioned close together, increasing the plates' size to offer more surface area, using a more powerful insulator between the conductive materials, using more conductive materials, etc.

CIRCUITS

Have you ever wondered how with the flip of a switch, lights, TVs, and computers can turn on? Flipping a switch completes an electrical circuit, which allows for current to travel and household appliances to work. In the simple case of a lightbulb, for electrons to produce light, they need to travel through the lightbulb and back, which is accomplished by a circuit.

The main task of an electric circuit is ensuring that the current can travel from one point to another. This current is pushed through a force known as voltage, which is produced through a battery or generator connected to the circuit. Circuits can transmit millions of volts worth of electricity over thousands of miles (powerlines) or through tiny microelectronic chips inside a computer. Therefore there are two main branches of electric circuits; power circuits and electronic circuits. Power circuits control and transmit megavolts of electricity. The major components of a power circuit are an electric generator or battery at one end, transformers and circuit breakers in between, finally, ending with some appliance at the end. Electric circuits, on the other hand, are more focused on processing and transmitting the information. These are used in computers, radars, cell phones, etc.

MAGNETISM

The reason for the existence of magnetic fields (or magnetism) is still unknown to a large degree (though there are theories). However, just like electric fields are produced by an electric charge, a magnetic field is produced by a magnet.

MAGNETIC FIELD

There are three straightforward ways to create a magnetic field. The first way is to use a magnet. The second way is by using current (the movement of electrically charged particles) to make an electromagnet. However, a current can produce a magnetic field that can be used for a variety of tasks. The third way is to use particles such as electrons because they have an intrinsic magnetic field around them; this is why a permanent magnet (e.g., the type used on refrigerators) has an endless magnetic field. Contrarily, there are times where other materials result in the cancelation of magnetic fields within electrons giving no net magnetic field around the material.

There are two poles of a magnetic field, the north, and the south. Like poles repel, and opposite poles attract. A critical note about magnets is that there is no way to independently simulate the North and South poles. For example, if one were to cut a magnet in half, there would be a creation of two separate magnets with two different poles. Now, if you were to put these two pieces back together, the new poles disappear.

Materials that commonly makeup magnets (iron, nickel, cobalt, etc.) are considered elementary magnets themselves, with an N and S pole each. Therefore, a magnet actually consists of many elementary magnets; this is why when a magnet is split, two new poles appear. The difference between a magnetized object and a non-magnetized object is the alignment of these tiny elementary magnets. The tiny elementary magnets inside are aligned arbitrarily in an unmagnetized object, causing them to be canceled out.

A magnetic field is also represented with field lines, similar to that of electric fields. At the points near the poles, the density of these magnetic field lines is high, while, near the center of the magnet, the density of these lines is low. Additionally, if one keeps a magnetic compass, it will realign itself in the same direction as the magnetic field lines. Hence, a compass always points north, no matter where you are in the world because the Earth can also be described as a giant magnet.

EARTH AND MAGNETISM

Earth is a magnet. Its magnetic field can be described as a huge bar magnet, where the field is strongest at the ends (e.g., the poles). At any point on Earth's surface, the magnitude and direction of the magnetic field would differ. The observed magnetic field also varies over the course of several years (and quite a lot over the course of 100 years). The true cause of Earth's magnetic pole is not understood, though it might have something to do with its electric field and Earth's inner core. Theories about the generation of Earth's electric field will be covered later in this book.

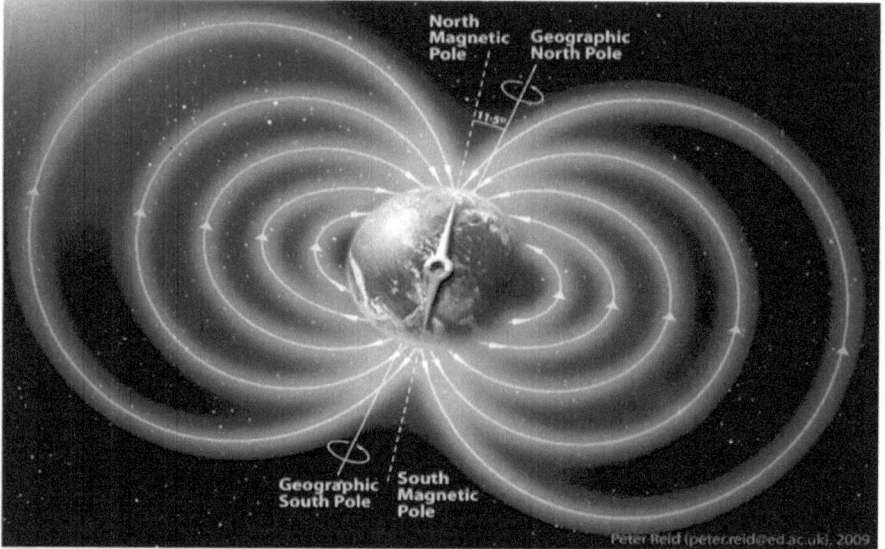

FIGURE 84: EARTH'S MAGNETIC FIELD [27]

ENERGY IN A MAGNETIC FIELD

Contrary to popular belief, magnetic fields do not emit energy. Additionally, after a magnetic field has been created and is static, there isn't a need to add a constant stream of energy to maintain such a field. Yet, the creation of the magnetic field would require a starting amount of energy.

Energy is still stored inside a magnetic field, and this principle allows for inductors to function. Like how a capacitor's energy is stored in the electric field between its plates, an inductor can also store energy in its magnetic field. The amount of energy stored varies depending on the size of the inductor and other variables; fields can be considered a means to store energy to be used later. The amount of energy stored varies depending on the size of the inductor and other variables.

ELECTROMAGNETISM

Electromagnetism is a branch of physics that addresses the interaction between electricity and magnetism. As we will discuss in this chapter, there is a special relationship between magnetic fields and electricity.

Additionally, the EM (Electromagnetic) spectrum is a crucial aspect of our lives, and it's used to power things ranging from motors to the human brain. We will discuss all of these aspects in a bit more detail in the following chapter.

LORENTZ FORCE

Named after the Dutch physicist Hendrik Antoon Lorentz, the Lorentz force defines a force that acts on moving charged particles (like electrons or protons) in a magnetic field. Suppose the magnetic field is oriented perpendicular to the direction of the electron's motion. In that case, the force experienced by the electron is perpendicular to both the direction of motion and orientation of the field. In other words, if there is a circular magnetic field all around an object and an electron is moving to the right, the Lorentz Force acts in the upward direction. The Lorentz force is commonly used to explain the Hall Effect.

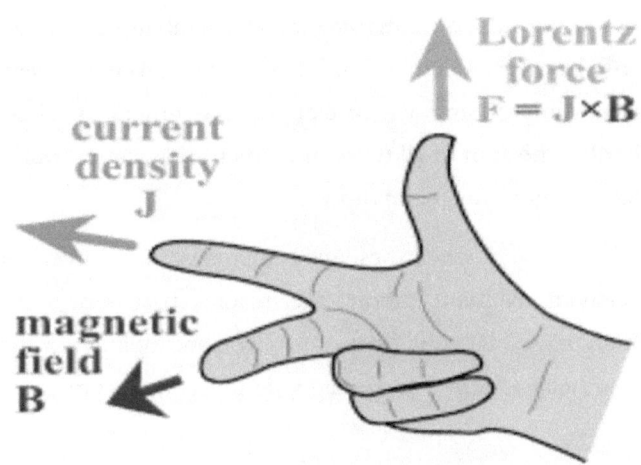

FIGURE 85: RIGHT-HAND RULE FOR LORENTZ FORCE [28]

HALL EFFECT

In 1897, when analyzing the current flowing in one direction through a conductor or semiconductor, physicist Edwin Hall noticed something peculiar. When moving through a conductor, such that there is a voltage, there is an accumulation of negative charge toward the conductor's bottom edge; this is because when electrons are flowing towards the left, through a conductor, the Lorentz Force is downwards instead of upwards (remaining perpendicular). This accumulation results in a voltage that develops between the upper and lower edge of the conductor, where the lower edge is more negative than the upper edge. This potential difference is known as the Hall voltage.

CURRENT AND MAGNETIC FIELD

A magnetic field is a tool used to describe how a magnetic force is distributed in the space around it. A straight wire carrying current results in creating a magnetic field; this is due to electrons inherently having a magnetic field around them.

MAGNETIC DIPOLE

A magnetic dipole is a miniscule magnet of subatomic dimensions. The movement of electrons, whether that be around an atomic nucleus, spinning on their axis, or any other form of electron movement, is all considered a magnetic dipole. The sum of all these movements can cancel out, resulting in a material having more magnetic dipole.

On the other hand, when these movements do not cancel, the atom can have a permanent magnetic dipole; this results in materials such as iron, which naturally have magnetic properties. In iron, millions of atoms knock against each other to form a ferromagnetic domain, a permanent magnetic dipole.

SOLENOID FIELD

A solenoid is a lengthy coil of wire winded tightly around a cylinder with a current passing through it. Within the coils, a strong magnetic field is created whenever a current is run through the wire—the direction of the magnetic field depends on the current direction. A solenoid acts like a permanent magnet that can be reversed and turned on and off, though this field is not perfectly uniform. Near the edge or outside of the coils, the magnetic strength is no longer consistent; still, this is a pretty ingenious mechanism since it provides a simple means to generate a strong magnetic field. Additionally, it can also be used to convert current into mechanical motion.

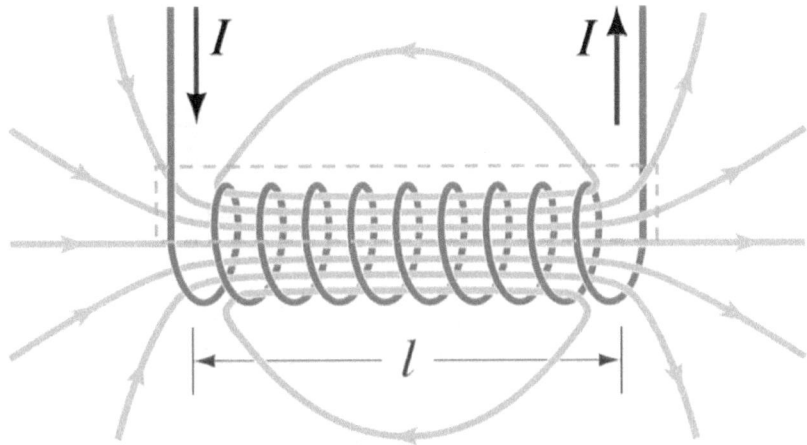

FIGURE 86: SOLENOID MAGNETIC FIELD [29]

Iron, cobalt, and nickel have powerful magnetic properties. Also, changing the material of the solenoid will result in either stronger or weaker magnetic fields. Generally, iron is commonly used as a household solenoid to create switches and turn lights on and off.

ELECTROMOTIVE FORCE (EMF)

It's known that in a generator or battery, energy is converted from one form to another, where one terminal of the device becomes positively charged, and another becomes negatively charged. The work (or transfer of energy) done on a unit of electrical charge is the electromotive force. This force is a characteristic of any energy source capable of driving an electrical charge around a circuit. Without this force, if one were to hook up a battery and a light build, the bulb would never light because the electrical current or charge would never travel to the bulb. There would be nothing to push the current through the wire to the bulb; this is why the electromotive force is necessary and ensures that the current spreads throughout a system.

But, the electromotive force isn't necessarily a force, despite its name. It's measured in volts of electrical charge; this is not equivalent to the Newton that is commonly used to measure forces.

INDUCTANCE

Inductance is caused due to the magnetic field generated by the electrical currents flowing through the circuit. It is also known as an inductor's ability to store energy through the magnetic field created via the flow of electrical current. There are two different ways inductance is used: self-inductance and mutual inductance.

Self-inductance is the property of a circuit (usually a coil) where when two coils are near each other, a current in one coil produces a magnetic flux through the second coil. On the other hand, mutual inductance is where the charge and current in one circuit results in charge and current in another due to a magnetic field that links both circuits.

Inductance and capacitance are commonly confused, but the main difference is that inductors store power as a magnetic charge while capacitors store electric charge. As stated previously, in a capacitor, opposite charges build up between two elements; this is different from an inductor where the current flowing through it builds the magnetic field.

DIRECT CURRENT

Direct current or DC is the flow of electric charge that doesn't change direction. Direct current (it was later supplanted by an alternating current or AC) is produced by batteries or fuel cells.

The transition to AC or alternating current was due to the issues that arose to transfer the high voltages needed for long-distance transmission. Later techniques were developed to circumvent this issue, and direct current is now transmitted over long distances, though it is consistently converted to alternating current for the final distribution.

ALTERNATING CURRENT

Most digital electronics use direct current, which only flows in one direction. Electric charge in an alternating current changes direction periodically. Current starts at zero, then grows to a maximum, decreases to zero, reverts reaches a maximum in the other direction, and returns to its original value. The circuit, then, repeats this cycle infinitely.

Alternating currents are commonly used for commercial and household products because they have a considerable advantage over direct currents. In a direct current, when trying to transmit electric current over large distances, a great deal of energy is lost due to resistance. This issue is not present in an alternating current, and energy can be transferred to large spaces without huge concerns.

Additionally, transmitting and using high currents in DC circuits is incredibly tricky. For this reason, until the mid-19th century, most power grids used incredibly low voltages. However, as alternating circuits began to be used, higher voltages were able to be supplemented. And, transformers could be used to change the electric charge, which was not previously possible. Alternating currents are the only reason that power grids can transfer thousands of volts of current, but at the same time provide 120 volts to individual customers in the USA.

ELECTROMAGNETIC RADIATION

Electromagnetic radiation or waves is a form of energy that is all around us. This radiation takes many different forms, including radio waves, gamma rays, X-rays, light waves, etc. Originally, electricity and magnetism were taught to be separate forces; later, upon Maxwell's discoveries, electromagnetism studies how electrically charged particles interact with each other in a magnetic field.

Electromagnetic radiation is created when an electric field accelerates an electron. This movement produces oscillating magnetic and electric fields. At these varying fields, ensure the propagation of waves at different frequencies and wavelengths. These waves have a constant velocity (299,792,458 m/s) within a vacuum. It was found that these waves shortened within transparent mediums, with their frequency remaining the same; this means that some mediums are entirely opaque to a specific range of frequencies of electromagnetic radiation.

Electromagnetic waves are made up of particles known as photons and can behave like particle waves. Therefore, they are classified into various categories depending on their frequencies. In general, the energy carried by these waves increases as the frequency increases, and the wavelength decreases. When the wavelength of EM waves is more than the upper-end UV (Ultraviolet) rays, they contain enough energy to ionize matter (which is quite a problem in outer space).

Electromagnetic radiations are transverse waves that have electric and magnetic components. This dualism makes them quite different from other waves such as sound, ocean, gravitational, or matter. Yet, they still have all of the same properties as other pressure waves, which leads to many phenomena being associated with them.

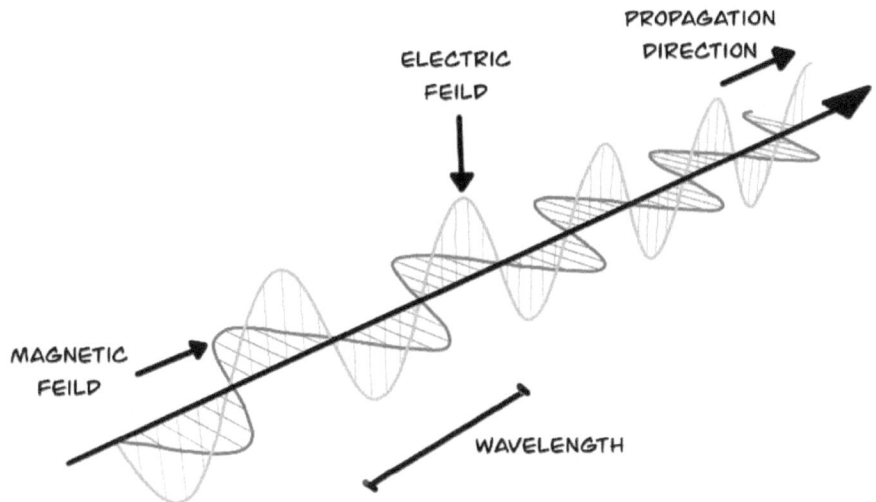

FIGURE 87: INTERNALS OF AN ELECTROMAGNETIC WAVE

MAXWELL'S EQUATION

If one spends enough time studying physics, Maxwell's equations will eventually come up. Since most of our day-to-day devices are primarily based on EM waves, this equation is incredibly important. These iconic equations are a big deal in physics since they model an electromagnetic wave (or light) or most of the radiations in the Universe. There are four clear equations, named after 19th-century physicist James Clerk Maxwell, which address the fundamentals of electromagnetic radiation. This book will describe the concept behind these equations without the intrusion of complicated vector calculus. These four laws or equations are:

1. Gauss' Law for Electric Fields
2. Gauss' Law for Magnetic Fields
3. Faraday's Law
4. Ampere-Maxwell Law

GAUSS' LAW FOR ELECTRIC FIELDS

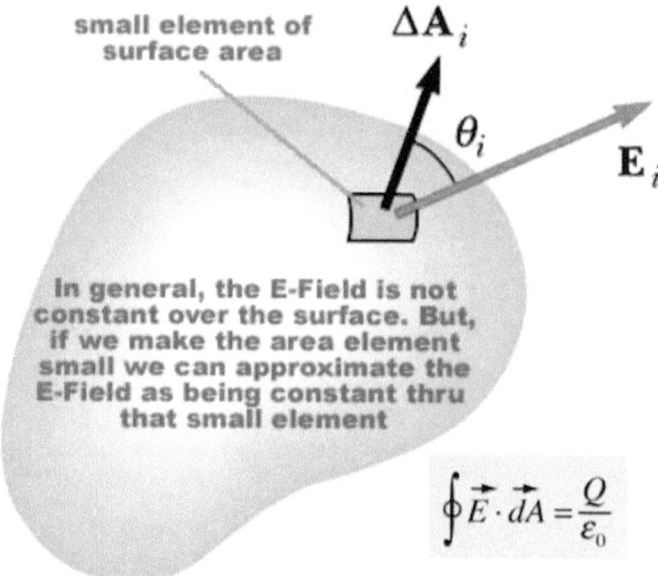

FIGURE 88: GAUSS' LAW FOR ELECTRIC FIELDS

First is Gauss's Law for Electric Fields, which describes the electric field pattern created due to electric charges (this law is also known as Maxwell's first law). Simply speaking, Gauss' Law states that electric fields point away from positive charges and toward negative charges (as stated previously). Additionally, electric fields follow the superposition principle, which means that the total electric field at any position is the vector sum of the electric field due to the point charges nearby.

In simpler terms, imagine an enclosed surface (like a sphere or a sealed box). The total strength of all the field lines of flux coming out of that box is directly proportional to the electric charge present inside of the box. If there is a positive and negative charge inside the box, the total electric field flux outside the box is zero (since the positive charge cancels out the negative charge). This equation discusses only the net charge and the total electric field flux of an enclosed surface. Note this equation concludes the following:

1. Electric charge produces an electric field.

2. The total flux of the electric field is proportional to the net electric charge present inside the enclosed surface.

It is perfectly okay if you don't understand the mathematical equation described in the image above. Note, in general; it goes over the concepts we stated previously. Since electric flux is dependent on the electric field inside the surface and the surface area, it is depicted on the equation's left side. It can be calculated as a dot product between the electric field vector and the surface area vector. Since this is an enclosed surface, a surface integral is taken. On the right side, the Q represents the net electric charge inside the body, and the e_0 is a constant (it describes the permittivity of free space, also known as a vacuum).

PERMITTIVITY OF THE MEDIUM

The permittivity value of a material depends on its response to the applied electric field. In nonconductors or insulators, the charges in the material do not move freely, and hence the permittivity value will be high. Also note, Gauss's Law for Electric Fields is valid for all materials provided we consider all the charges present in the material. This permittivity value is particularly important in determining the speed of propagating electromagnetic waves within that medium, which you will learn about in later sections.

GAUSS' LAW FOR MAGNETIC FIELDS

Also known as Maxwell's second law, though the equations associated with the law look similar to the previous one, they, instead, address magnetic fields. In contrast, the previous one addresses electric fields. Additionally, this law came after the previous one, hence the arbitrary naming. It's similar to if there is a student named Bob in a class, but another kid joins the course and is also called Bob. The class would call this student, Bob 2.

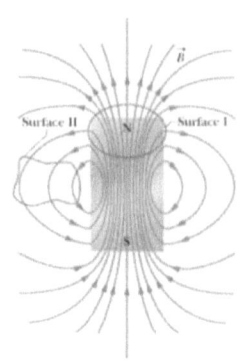 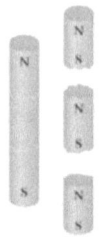

The simplest magnetic structure that can exist is a magnetic dipole. Magnetic monopoles do not exist (as far as we know).

Gauss' law for magnetic fields is a formal way of saying that magnetic monopoles do not exist. The law asserts that the net magnetic flux Φ_B through any closed Gaussian surface is zero:

$$\Phi_B = \oint \vec{B} \cdot d\vec{A} = 0$$

FIGURE 89: GAUSS' LAW FOR MAGNETIC FIELDS

There is a unique similarity between an electric field's shape to a dipole and a magnetic field. The reason they look similar is that, mathematically, they are the same. Additionally, as stated in a previous chapter, if one were to break a magnet, there are now two separate magnetic fields instead of one individual magnetic pole (e.g., a magnetic monopole does not seem to exist). This idea is inherently what Gauss' law for magnetism states - there is no such thing as a magnetic monopole or has never been witnessed.

Simply put, again, think of any real or imaginary enclosed surface (e.g., like a sphere or a closed box) the total strength of all the magnetic field lines or the flux coming out of that surface always is zero as there is no monopole. Since there will always be north and south pole pairs, the total magnetic field flux in an enclosed surface is still zero (since the north pole of a magnet cancels out the south pole).

In the equation depicted, the B represents the magnetic field vector, and its unit of measurement is a Tesla (named after the famous physicist and inventor Nikola Tesla). The magnetic flux is measured in Webbers. The Gauss equation for magnetism is similar to the electric field law; however, it is equal to zero on the right side. It is assumed that no known magnetic monopole exists (though in 2009 in Germany, under special conditions, they found a magnetic monopole to exist).

FARADAY'S LAW

Directly put, Faraday's Law (also known as Maxwell's third law.) explains that electric charges are not the only way to make electric fields. One can also make an electric field by manipulating a magnetic field; this is the central idea that creates the bridge between electricity and magnetism. A classic demonstration is used by moving a magnet in or out of a coil. It would show a current when connected to a galvanometer, thereby establishing a magnetic field producing electricity.

Note, a fundamental concept of Faraday's Law is that for there to be an electric field; the magnetic field needs to be changing. If one were to hold a magnet in the coil, there would be no current; this magnet needs to be moving. Additionally, this electric field makes circles instead of the classic magnetic field described earlier (pointing away from the positive and negative charge). This is caused by the changing magnetic field, which doesn't exist at a fixed point.

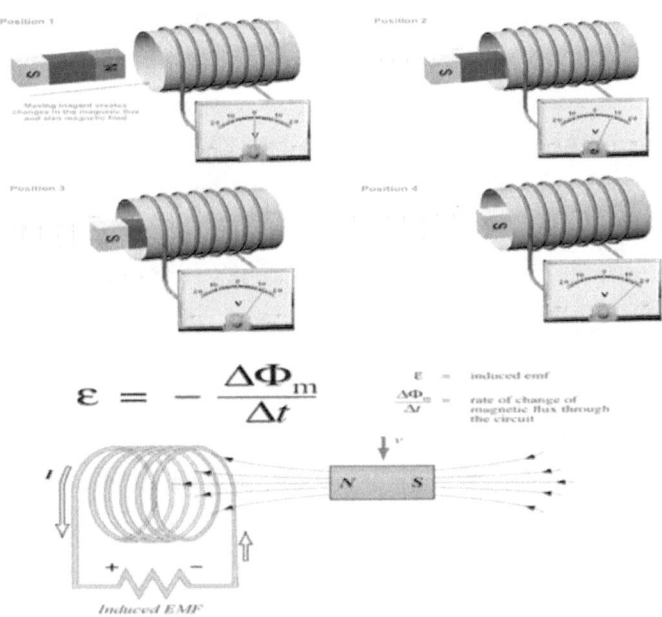

FIGURE 90: FARADAY'S LAW EXPLANATION

Note Faraday's law is key to the functioning of devices such as generators. As we know, an electric generator works by rotating magnets around a coil that produces electricity due to changing magnetic fields. Changing magnetic flux produces an electromotive force on the boundary of the surface. This induced EMF drives the current through the material, thereby generating current.

In short, changing a magnetic field induces a circulating electric field, and changing magnetic flux produces EMF. Also, note that induced EMF opposes the change in the magnetic flux (Lenz's law). There is one difference between the electric field produced by an electric charge and one produced by a changing magnetic flux. In a magnetic flux, the electric field lines created loopback. This form of electric field lines will have no point of origination and no point of termination. In electric fields produced by an electric charge, they originate from a positive charge and terminate in a negative charge.

LENZ'S LAW

Lenz's law is an extension of Faraday's law (described later) when discussing electromagnetism. Named after Russian physicist Emil Lenz, the law states that an induced electromagnetic force with different polarities generates a current (similar to Faraday's Law). However, the direction of this current is always opposite to the charge of the magnetic field that produces it; this ensures that the original flux is maintained through the loop while the current flows in it. Even the formula for Lenz's Law reflects Faraday's law, showing their parallelism.

AMPERE-MAXWELL LAW

This law is also known as Maxwell's fourth law. This equation looks similar to Faraday's law, thereby hinting at parallels in how electricity and magnetism work. There are two ways to make an electric field, and there are also two ways to make a magnetic field. A standard demonstration would be having a battery and a wire connected to it. If one were to put a magnetic compass under the wire when the electric current flows, it creates a magnetic field, shown with the moving compass needle.

Additionally, the parallels continue, the same way the electric field makes a circle, this magnetic field also creates circles, which look similar to coils. In layman's terms, the magnetic field is "curly." Again, changing the electrical field resulting in a circular magnetic field. In summary, changing electric fields make circular magnetic fields, and changing magnetic fields make circular electric fields.

ELECTROMAGNETIC WAVES

Suppose one has a region of space with nothing but an electric field and a magnetic field; there is no electric current. Even with the lack of a current, a magnetic field is created under Faraday's Law, and an electric field can be created due to the magnetic field. An electromagnetic (EM) wave is created with an oscillating electric field and an oscillating magnetic field in the perpendicular direction.

Most waves need a medium to travel; however, due to the unique phenomenon that create EM waves, they can travel through a vacuum. The interconnection between magnetic and electric fields ensures that an EM wave propagates perpendicular to both the electric and magnetic fields. Since it can travel through a vacuum, it can also travel through space, allowing people on Earth to get light (also an EM wave) from the Sun.

RADIATION PRESSURE

Radiation pressure is pressure on a surface resulting from EM waves. This pressure is a result of the momentum carried by that radiation. Although the pressure from solar radiation is minimal, a sufficiently large surface would produce a technically useful force, such as creating solar sails in the movement of spacecrafts.

Radiation pressure is also commonly viewed in space. Most stars inhabit a bit of gas pressure in space due to the sheer amount of photons they emit. Depending on the star's mass, this gas pressure created due to the radiation pressure begins to dominate and sets the upper limit for how massive a star can become. The pressure from solar photons is also partially responsible for the creation of dust tails behind comets. Additionally, the formation of planetary nebulae is also primarily based on radiation pressure. As a star dies, it releases vast amounts of heat and light. The radiation pressure given off is so strong that the star's outer layers are pushed out and form a gaseous nebula surrounding the star.

ELECTROMAGNETIC SPECTRUM

A beam of light is traveling as a wave of electric and magnetic fields (e.g., an electromagnetic wave). Electromagnetic waves are such a common phenomenon that a huge spectrum is used to identify and describe the different aspects of this wave. These waves can be classified and arranged based on their varying frequencies and wavelengths. All electromagnetic waves, no matter where they lie on the spectrum, can travel through a vacuum.

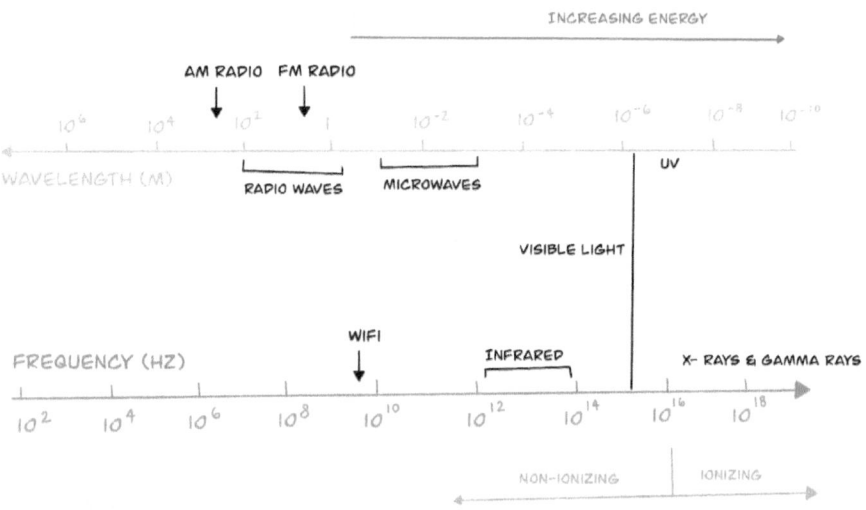

FIGURE 91: ELECTROMAGNETIC WAVE SPECTRUM

One of the most profound regions of the EM spectrum is visible light. These waves have a relative sensitivity to the human eye. The center of the visible region produces a sensation that we call yellow-green. The limits to the visible spectrum are not necessarily very well defined. Still, it is between the wavelengths 430 nm and 690 nm; note, the eye can also detect other electromagnetic waves if they are intense.

VISIBLE LIGHT

Visible light is a form of electromagnetic radiation, visible to most human eyes. A unique characteristic of visible light is color. Color is both a property of light and a fabrication of the human eye. Objects don't inherently "have" color; they give off light that "appears" to be a color. Therefore color only exists in the person who sees the object.

Human eyes contain special cells known as cones, which act as receivers to the wavelengths. Light at the lower end of the spectrum has longer wavelengths and is viewed as red. Light in the center of the spectrum is viewed as green, and light at the upper end is seen as violet. All other colors are mixtures of these colors. For example, yellow contains both red and green; cyan is a mixture of green and blue. Isaac Newton was the first person who realized that white light is made up of all the colors in the rainbow through a prism experiment where white light was passed through a narrow slit, and the color spectrum is projected on the wall.

As objects grow hotter, they radiate light in shorter wavelengths, which humans see as changing colors. For example, when one turns on a blowtorch, the color changes from orange to blue if adjusted to burn hotter. The property of changing thermal energy into light is called incandescence. Incandescent light is produced when a hot object releases its thermal vibrational energy as photons. As temperature increases, the energy moves into the visible light spectrum and appears to have a reddish glow; over time, and eventually to blue.

Astronomers can determine the makeup of specific objects because each element absorbs light at particular wavelengths. By determining what elements are absorbed, astronomers can use spectrographs to determine the chemical composition of stars. The color of hot objects can be used to estimate their temperatures. For example, the sun's surface temperature is assumed to be emitting light at the peak wavelength of slightly yellow.

INFRARED WAVES

In the 1800s, British astronomer William Herschel discovered infrared light in an experiment measuring the temperature difference between colors on the visible light spectrum. He observed an increase in temperature from blue to red and an even warmer temperature just beyond red at the end of the spectrum. Infrared waves occur at frequencies above microwaves and below red light (hence the name 'infrared').

Infrared radiation is a type of radiant energy invisible to the human eye; however, it is felt in the form of heat. All objects in the universe emit infrared or thermal radiation, the most prominent ones being the Sun and fire. Infrared radiation is created through the frequencies produced when atoms absorb and release energy.

Like visible light, infrared also has its range of wavelengths. The shorter waves are closer to visible light and don't emit any detectable heat. The longer waves are closer to the microwave section and can be felt as intense heat, such as a fire. This form of radiation is one of the ways that heat is transferred from one place to another.

ULTRAVIOLET RAYS

Ultraviolet rays or UV light is a type of electromagnetic radiation that makes black-light posters glow and causes skin cancer (if exposed in extreme amounts). Electromagnetic radiation comes from the sun and is transmitted at different wavelengths and frequencies. These wavelengths are known as the EM spectrum, and UV rays are one aspect of this spectrum.

Interestingly, UV radiation has enough energy to break chemical bonds. The higher energies can cause ionization, a process in which the electrons break away from atoms. This vacancy of electrons affects the chemical properties of the atoms causing them to form or break chemical bonds. It's this property that can make UV radiation harmful to living tissues. Some damage can be beneficial, such as disinfecting surfaces. Still, this can also harm biological systems such as skin and eyes, which is why sunscreen is always recommended on a beach.

RADIO WAVES

The building block of radio communications is the radio wave. Radio waves are a form of electromagnetic or EM radiation. They also have the lowest frequencies compared to all the other waves allowing for easy use in communication. Depending on their use, radio waves are divided into nine different bands for communication purposes.

Radio waves are also commonly used to determine the composition of cosmic objects. A radio telescope picks up distant pulsars or quasars. These are created with extremely powerful cosmic events sending waves of energy and radiation to Earth. Additionally, these are the large series of dishes that one sees when viewing an astronomy outpost.

AMPLITUDE MODULATION (AM)

Amplitude modulation is a technique used for electronic communication by combining multiple waves. As described by the name, amplitude modulation focuses on radio communication by changing the wave's amplitude. For information to be transmitted via a wave, it needs to be modulated or changed somehow. In amplitude modulation, two waves are used; one is a carrier wave, and the other is the signal. The signal contains the information to be transmitted. Generally, the carrier wave will have a much higher frequency than the signal. These two wave amplitudes are added directly to generate the wave which will be transmitted. In radio transmission, sound waves (voice and music) are mixed with higher frequency radio waves (electromagnetic waves), and the antenna transmits the modified carrier wave.

This form of amplitude modulation is the easiest and simplest to use because it doesn't require complicated demodulators.

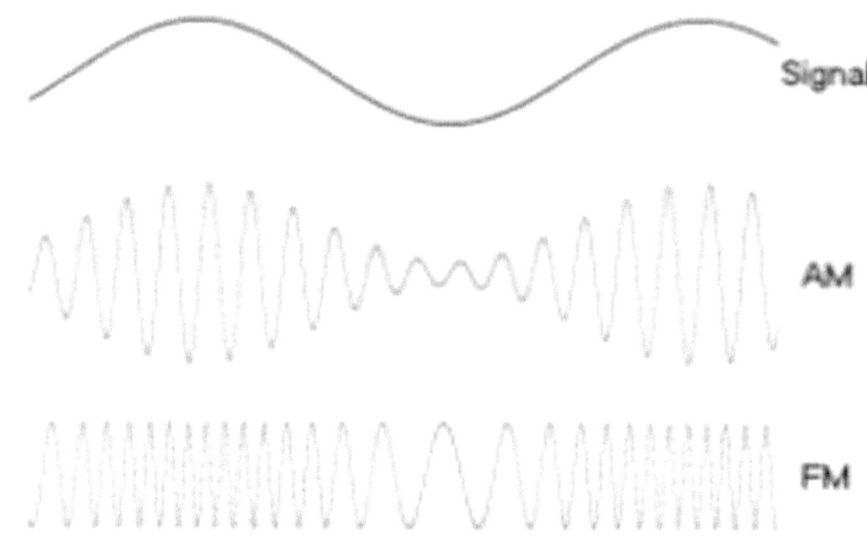

FIGURE 92: AMPLITUDE AND FREQUENCY MODULATION [30]

FREQUENCY MODULATION (FM)

Frequency modulation relies on changing the frequency of the carrier wave for transmitting the signal. In frequency modulation, two waves are also used. One is a carrier wave, and another one is the signal. The signal contains the information to be transmitted. Generally, the carrier wave will have a much higher frequency than the signal. In this technique, the instantaneous frequency of the carrier wave is modified to transmit the signal. Similar to amplitude modulation in radio transmission, sound waves (voice and music) are mixed with higher frequency radio waves (electromagnetic waves). The modified carrier wave using frequency modulation is transmitted by the antenna.

ANTENNAS

Antennas are the basis of how radio signals are transmitted throughout the world. Microphones are used to capture sounds, and they are turned into electrical energy. This electricity flows through tall metal antennas, which boosts its power several times over. Then electrons vibrate back and forth along with the antenna, thereby creating EM radiation through radio waves.

These waves travel at the speed of light, taking the sound or information with them, allowing you to hear them once you turn on your radio.

The simplest antenna is a piece of metal attached to a radio. Most modern antennas have two different radios, one being able to catch AM signals and the other to detect FM signals; this is because of the varying wavelengths between these two forms of transmission. In this way, the antenna's main job is to pick up enough energy from radio waves to make the circuits resonate at a specific frequency.

There are multiple ways that radio waves can travel throughout the world. The primary way radio waves travel is by shooting up to the sky and bouncing off the ionosphere, and traveling back to the ground; this allows for most radio communication on Earth. Without the ionosphere, it would be difficult to transmit radio waves over long distances (Marconi was lucky there is an ionosphere to reflect the radio waves from Europe to the USA). Hence, radio waves are easier to pick up in the evening because they have a more challenging time bouncing and traveling in the mornings due to the ionosphere's general structure.

MICROWAVES

Microwaves are a section of the electromagnetic spectrum with frequencies ranging from about 1 billion Hz to about 300 billion Hz. These waves are most commonly used for point-to-point communications to convey information ranging from voice to video in analog and digital formats. These waves are also used in 'radar.' Radar is, in fact, an acronym for Radio Detection And Ranging. Before WWII, British radio engineers found that radio waves could be bounced off distant objects, and the returning signal could be detected with sensitive antennas. Thereby giving individuals the locations of these objects, and were further used in submarines and ships to navigate at sea.

One commonly uses microwaves to warm up food quickly. The invention of the microwave oven and the discovery of this phenomenon was made purely by accident. After WWII, electrician Percy L. Spencer while touring one of his laboratories, stopped in front of a magnetron (the power tube that drives the radar set). He notices that the candy bar in his pocket began to melt, leading to the furthering of this invention. The reason microwaves work is because these waves cause water molecules to vibrate. The increased friction between the molecules results in heat, so it isn't entirely safe to be around highly powerful microwaves.

Scientists have also observed extraterrestrial microwaves coming from deep space. In 1965, two Bell Labs scientists detected background noise using long L-band microwaves with a special low-noise antenna. Interestingly, the noise seemed to be coming from every direction and didn't vary much in intensity. Researchers soon realized that they had discovered the cosmic microwave background radiation, which fills the entire universe and a clue to the very beginning.

X-RAYS

X-rays were discovered by Wilhelm Conrad Röentgen in 1895 when experimenting with cathode-ray tubes. Röentgen noticed that crystals in a high voltage tube exhibited a fluorescent glow, even when he covered them with dark paper. Therefore, the tube produced some form of energy, which was penetrating, causing the crystals to glow. Röentgen coined this energy, "X-radiation." Later experiments showed that this radiation could penetrate soft tissues, but not bone, and produce shadow images on photographic plates.

X-rays can be produced by sending a high-energy beam of electrons smashing into an atom, such as a copper. When the beam hits the atom, the electrons in the inner shell get jostled and flung out of their orbit. Without those electrons, the atom becomes unstable. The electron from a higher energy level drops down to fill the gap, and an X-ray is released to stabilize itself. However, X-rays' main issue is that they are not directional or focusable, leading to x-ray machines being challenging to focus on.

GAMMA RAYS

Gamma-rays were first seen by French chemist Paul Villard when he was observing radiation from radium. A few years later, Ernest Rutherford proposed the name 'gamma rays' when investigating alpha and beta decay. Since gamma rays are emitted during alpha and beta decay, the name that was given stuck. Gamma rays are commonly witnessed during four different nuclear reactions: fusion, fission, alpha, and beta decay.

Gamma-rays can overlap with X-rays making it challenging to differentiate between them. Usually, an arbitrary line is drawn in the EM spectrum where the rays above a particular wavelength are classified as gamma rays. Gamma-rays also have enough power to harm living tissue.

Thankfully, almost all cosmic gamma-rays are blocked by Earth's atmosphere.

LIGHT AND OTHER FORMS OF ENERGY

Everyone is familiar with the famous scientific experiment conducted by Isaac Newton via a glass pyramid. If you've never heard of this, then consider what you have heard about white light and the rainbow in your high school physics class. It has been known that white light from the Sun has a wide variety of wavelengths and frequencies, thereby creating white light. White light is the combination of all the rainbow seven colors (this was discovered by Sir Isaac Newton when he refracted sunlight through a glass pyramid). He found that the light that started as a single white beam split into many different colors after hitting and exiting the pyramid—thereby making a beautiful rainbow on the other side.

What Newton did all those years ago was the first example of geometric optics. Optics is a branch of physics that studies light along with all of its different behaviors and properties. Optics allows us to make lasers, telescopes, and many technologies that change the world. Therefore, in this section, we will be discussing the basics of optics, so let's get started.

LIGHT AS A WAVE

The first wave theory developed for light was Huygens's principle (named after Dutch physicist Christian Huygens). Huygens's principle is commonly used to explain the laws of refraction and reflection.

REFLECTION

Several different principles and laws govern light, one being the law of rectilinear propagation. This law states that light travels in a straight line, and when a ray of light encounters a surface (e.g., a transparent boundary) such as glass, the ray changes direction and travels along a new path. This motion of the light ray results in shadows, images, etc.

Additionally, the idea of a parallactic displacement or parallax is also important when addressing reflection. For instance, when you're looking at an object and close one of your eyes, it changes position than if you had both of your eyes open. This concept is known as parallactic displacement, and it can be used to find distances based on the location of specific objects.

Smooth surfaces reflect light incredibly well since a smooth surface does not change its direction rapidly; this ensures that the light rays also don't change their path quickly. On the contrary, surfaces that aren't smooth reflect light in several different directions. The definition of a smooth surface is based on the irregularities on the surface being at a minimum compared to the light's wavelength; this means that the light doesn't change a large amount from its original path.

REFRACTION

A typical example of refraction occurs when one views an object kept in water from above. The item seems to be shifted up; this occurs due to refraction at the water's surface; therefore, the real depth varies from the apparent depth. Refraction is a phenomenon that refers to the change in the direction of a wave passing from one medium to another. In the example above, the two mediums would be air and the surface of the water. Light travels from air into the surface of the water and hits the object, and bounces off. However, when the light wave bounces off the object, there is a shift when it enters the new medium, resulting in a shift of direction for the viewer.

When more than one medium is separating the object and the observer, a normal shift occurs. A clear example of a normal shift is when one were to look at a straw in a drink through the glass. The straw seems to have shifted from its position by some distance. This distance is known as the normal shift. The normal shift is only applicable if you're looking at an object through some transparent slab (e.g., the surface of a glass).

Refraction is also by which a prism splits white light into its different components. If one were ever to shine a flashlight into a prism, the output is a rainbow. When light enters and leaves the prism, the ray is bent downward; this is one of the main uses of a prism: the deviation of light from its path. This phenomenon also occurs inside a raindrop when sunlight hits it, resulting in a rainbow.

The material mediums which have refractive properties are classified using their refractive index. The formula that measures the Refractive index is:

Refractive Index = Velocity of light in vacuum / Velocity of light in the medium

INTERFERENCE AND DIFFRACTION

Everyone has witnessed dropping a pebble into a lake. The ripples or water waves start small but spread out as time passes. This simple phenomenon is a perfect example of diffraction. Diffraction is the tendency of a wave to travel through a passage and spread out as it propagates. In terms of light waves, this occurs when it reaches a whole or a tiny barrier. If you were to flash a flashlight down a dark tunnel, the light beam would spread out. And some areas would be dark while others would be light; this pattern is known as a diffraction pattern.

To better understand a diffraction pattern, we should address single slit diffraction. Let's say that you have two walls parallel to one another. We punched out an area between the two walls; the slit isn't too large; still, it isn't too small. If you were to shoot light through the hole, one would see that some areas are brightly lit, but others are dark. This mixture of light and dark is a clear description of a diffraction pattern. Note, to further understand why such a pattern is created; we need to discuss interference.

Everyone has heard of interference, specifically in the forms of communication. Usually, interference results in a loss of signal; but, what exactly is interference? Similar to radio waves, light waves are also a type of wave. They both have a wavelength and amplitude ranging in size. Wave interference occurs when two waves come close to one another. Constructive interference occurs when the crests of waves line up, thereby increasing the amplitude of the wave. The same is true if the troughs of a wave line up. Constructive interference leads to a more substantial wave that is either louder or more powerful. It is commonly used in modern technology to increase the amplitude of a wave.

On the other hand, if two waves are not aligned, and when a crest of one wave meets another's trough, one wave is dragged down by the other. The combined wave has shorter crests than the original wave or lower amplitude. This change within the wave is known as destructive interference. Interestingly if the two waves are shifted by exactly half a wavelength, the crests and troughs will match up perfectly, resulting in a straight line.

In light waves, destructive interference results in no light in a specific area or a dark spot. Constructive interference causes a specific spot to be brightly lit up. When two light waves enter a specific gap, there will be a diffraction pattern created because some areas will be dark due to destructive interference between the two waves. Other regions will be brightly lit up due to constructive interference.

COHERENCE

Coherence is a property of laser light and occurs from the stimulated emission process, which produces the light emitted from a laser. In general, coherence describes a phenomenon that occurs when emitted photons are "in step" and have a definite phase relationship with one another. This same principle is used to produce holograms.

Ordinary light is not coherent since it comes from independent atoms that emit light at different energy levels. There is a general coherence depending on the sources of light and other useful spectral sources. Yet, none of these are as coherent as the light from a laser.

LASERS

A laser (Light Amplification by Stimulated Emission of Radiation) occurs when the electrons in atoms, specifically in special glasses, crystals, or gases, absorb energy from an electrical current. This phenomenon results in the electrons becoming "excited." These excited electrons move from a lower-energy orbit to a higher-energy orbit around the nucleus of the atom. When these electrons lose energy and return to their normal or grounded state, the released energy is emitted as photons.

Interestingly, all these photons have the same wavelength and are therefore "coherent," coherence meaning that the crests and troughs of the light waves are uniform. This light also contains only one wavelength of a specific color determined by the energy released when the excited electron drops to a lower orbit. Laser light also has direction, and since it is coherent, this light can stay focused for vast distances depending on how powerful the laser is.

In large lasers, flashes of white light from giant flashlamps "pump" electrons in vast numbers into a slab of laser glass to a high energy state. The light emitted when these electrons drop to a low energy state is trapped in an optical switch. Mirrors at both ends of the laser amplify the photons through the glass, stimulating more and more electron drops. This process produces a considerable number of coherent photons amplified to create tightly focused laser beams that converge to make a large beam in the center of the laser.

CHEMICAL ENERGY

Molecules are groups of the same or different types of atoms that form other substances or compounds. The sharing of electrons between atoms forms a molecule. This communal exchange of electrons are also known as chemical bonds, and chemical energy is the energy stored in these bonds. This energy is commonly released when a chemical reaction takes place. For example, a battery is a form of chemical energy that is transferred into electrical power. In a battery, there are three main components: a cathode, an electrolyte, and an anode. In an anode, commonly in a battery, some chemical reaction occurs, which produces electrons. Simultaneously, in the cathode, a chemical reaction occurs to ensure that it can accept electrons from the anode; this creates a flow of electrons, thereby creating an electrical current and a charge.

A battery is a major way that chemical energy is harnessed to do work. There are two main types of chemical reactions that consume or release chemical energy: exothermic and endothermic reactions. An exothermic reaction is a type of chemical reaction that commonly releases energy in some form (e.g., heat, light, etc.). For example, an object burning is an exothermic reaction because it releases energy as heat and light. An endothermic reaction is a type of chemical reaction that occurs when there is energy absorption. For example, cooking an egg is an endothermic reaction because it absorbs heat from the pan to cook.

Commonly, for everyday purposes, humans use exothermic reactions to create and use chemical energy. Combustion is a common exothermic reaction used to produce energy from items such as coal or fossil fuels.

NUCLEAR ENERGY

Atoms are tiny units that make up the entirety of the universe, and there is a large amount of energy that holds the nucleus together (also known as the strong force). An atom consists of a nucleus at the center, and electrons revolve around it in different orbits. The nucleus of an atom consists of protons, neutrons, and various other atomic particles. Nuclear energy is a form of energy production derived from the nucleus, or core, of an atom.

There are two main ways nuclear energy is released; fission and fusion. In nuclear fission, atoms are split to release the energy housed in their nucleus. In fusion, atoms are combined to form a new kind of atom and release large amounts of energy. Commonly fission is used to produce energy since fusion requires very high temperatures and large amounts of energy. To produce electricity, nuclear reactors typically use uranium pellets as fuel.

These atoms split (fission), and they create a controlled chain reaction resulting in energy being released in the form of heat. If the chain reaction is uncontrolled, then the result will be an atom bomb or fission bomb.

This heat inside a reactor is used to warm a cooling agent such as water, which turns to steam. This steam then turns turbines, thereby creating a flow of electrical current and producing electricity. In the modern world, nuclear energy is considered to be the most reliable form of renewable energy. As of 2011, approximately 15% of the world's electricity is produced by nuclear power plants.

APPLICATIONS AND INTERESTING DISCOVERIES

Physics is always considered to be mainly in sync with engineering. Since many important discoveries made in the conceptual realm have profound impacts on technologies that can be created, specific laws and equations can be taken to build and create incredibly ingenious and mind-blowing devices. This chapter will briefly cover important physical discoveries made through engineering based on important physics principles. So, let's start.

SOME APPLICATIONS

Physics as a topic itself is quite impressive, and many physical concepts are commonly used in engineering. Inventors and scientists have taken advantage of interesting physical phenomena or laws to create technology that has dramatically improved humanity's quality of life. Here, in this chapter, we hope to describe the physical concepts behind some everyday technological creations probably located in your own home. This is a clear example of using physics in the real world and practical applications of universal concepts.

MOTOR

The general principle of an electric motor is straightforward. One puts electricity at one end, and an axel (metal rod) rotates on the other end. This action occurs due to the link between electricity, magnetism, and movement. Suppose you take a copper wire, loop it, and lay it between the poles of a powerful magnet. If you were to connect a battery to the two ends of the wire, the wire would briefly jump up; this happens because when a current is introduced within the wire, a magnetic field is created around it. Then the magnet interacts with the magnetic field, resulting in the wire jumping.

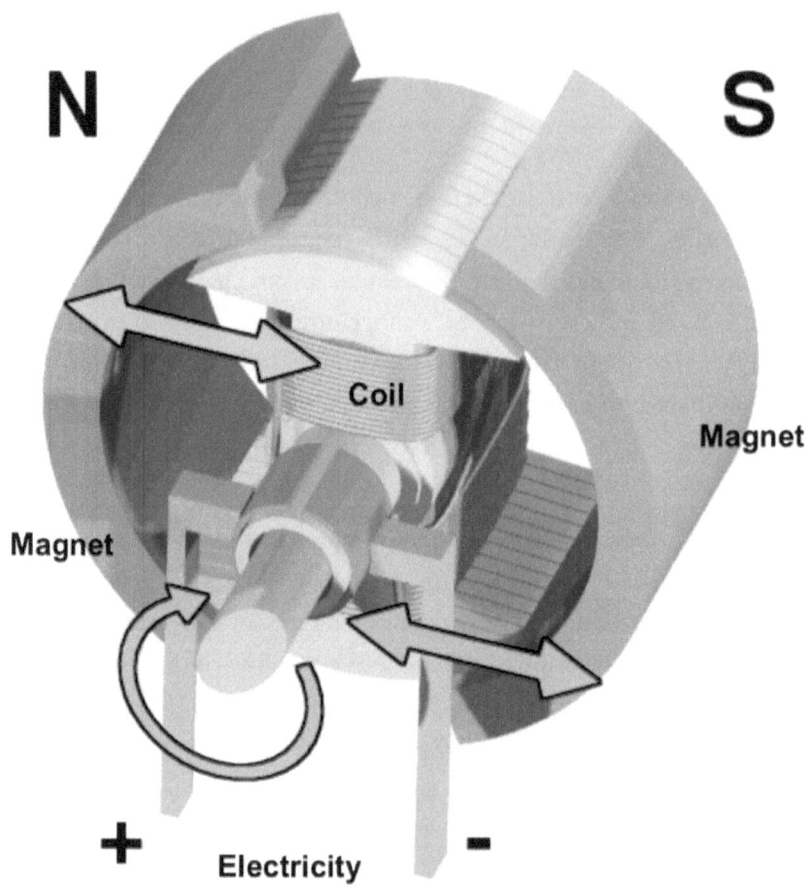

FIGURE 93: MOTOR INTERNALS [31]

Though there are various electric motors, the most common forms are universal electric motors that use an electromagnet. This electromagnet takes power from the DC or AC power that you feed into it.

When you feed in a DC current, the electromagnet works similar to that of a permanent magnet and produces an electric field that is always pointing in the same direction. When you feed in an AC current, the current flowing through the electromagnet and the current flowing through the wire both reverse. Therefore, the coil's force is always in the same direction, and the motor either spins clockwise or counterclockwise.

TURBINE

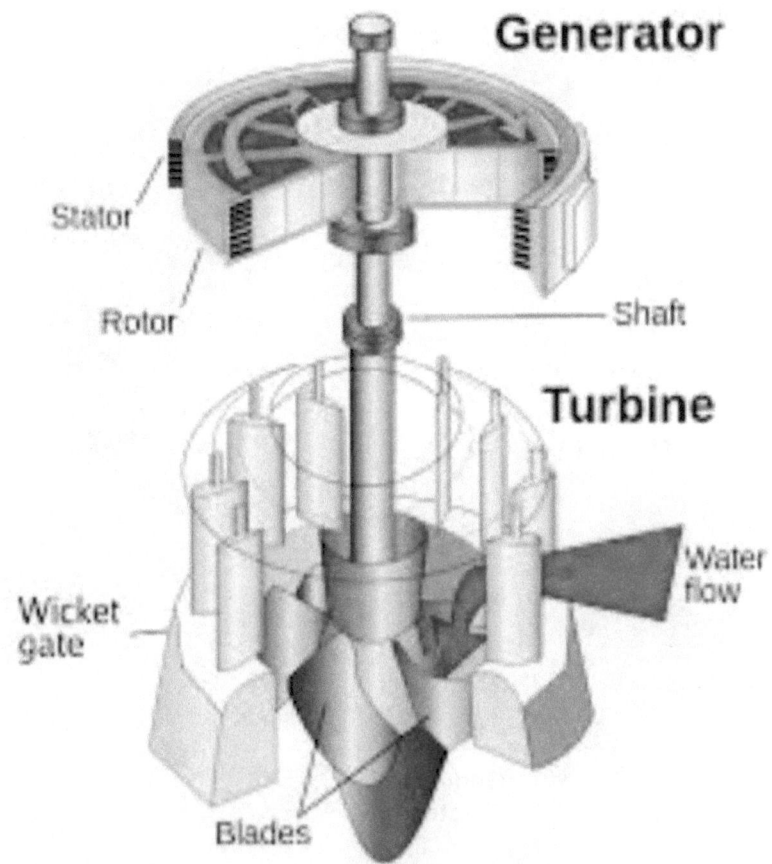

FIGURE 94: HYDRAULIC TURBINE [32]

Turbines are common aspects of modern technology. The most famous are those of wind turbines, which are used to generate renewable energy. Yet, turbines are also used within jet engines to hydraulic plants.

The mechanics of a turbine are relatively simple. It turns kinetic energy into electricity. The rotors spin around, capturing kinetic energy from the wind and turning the central drive shaft, which is connected to them. Although the rotors move quite fast, this central shaft moves much slower. Inside the turbine's main body, which is located behind the rotors, the gearbox transforms the shaft's low-speed rotation into a high-speed rotation fast enough to generate electricity through a generator. Finally, a transformer converts electricity into a much higher voltage to be easily transferred to a power grid, allowing homes to enjoy green energy.

GENERATOR

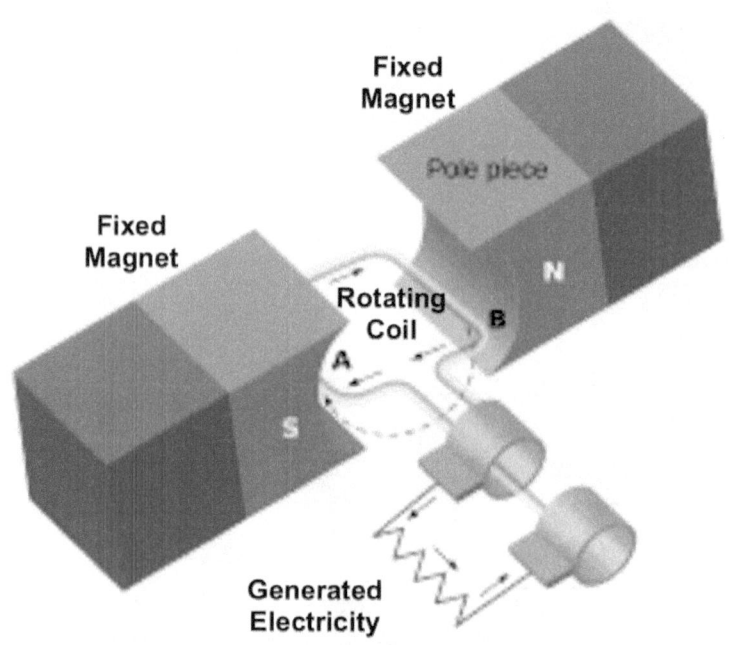

FIGURE 95: GENERATOR INTERNALS [33]

A generator is a simplistic device that moves a magnet near a wire to ensure a steady flow of electrons. The best way to think about a generator is a pump pushing water through a pipe. Instead of pushing water, the generator pushes electrons. Instead of pushing water, note, the generator pushes electrons.

A necessary clarification to make is that a generator doesn't "create" electrical energy. It converts mechanical energy provided by a force or torque into electrical energy. In the simplest terms, generators are a giant electromagnet and function using Faraday's Principle of Electromagnetic Induction.

Faraday found that electric charges could be created when a conductor moves in a magnetic field, which could be used to direct current through a wire. Generators force electrons from an external source through an electric circuit. Some electrical generators, such as the Hoover Dam, use water and transfer enormous amounts of power. Smaller household generators use fuel sources such as natural gas or propane to complete the job.

MICROWAVE OVEN

Similar to many great discoveries, the microwave oven was discovered by accident. Yet, by now, they have become a massive part of modern society. The way that these machines work is that there is a magnetron within the metal box that takes electricity and converts it into high-powered radio waves. Then the magnetron blasts these waves into the food through something known as a wave channel.

To ensure that the food inside cooks evenly, the plate of food rotates within the microwave. The microwaves bounce back and forth, reflecting off the walls of the container. Note, once the microwave contacts the food, it doesn't bounce off, thereby penetrating and heating it. Due to their large amount of energy, they can make the molecules vibrate more quickly by passing energy onto the food molecules.

Even though microwaves have a lot of energy, they cannot penetrate incredibly deep into the food. In fact, upon contact with food, they lose most of the energy they might have had initially; this is why one cannot cook meats in the microwave because a portion of it remains raw.

INTERNAL COMBUSTION ENGINE

Internal Combustion engines have been used since 1903 after the Wright brothers discovered their use when flying the first airplane. Today, these machines are used in almost all vehicles and aircrafts. The basis of an internal combustion engine is taking thermal energy and converting that into mechanical energy.

Every car uses a four-stroke combustion cycle within the internal combustion engine to convert gasoline into motion using the following four strokes:

1. Intake Stroke

    During the intake stroke, the piston begins by opening the intake valve and lets the engine take a cylinder full of gas and air.

2. Compression Stroke

    The piston moves to compress the gas and air in the cylinder, thereby making the explosion stronger.

3. Combustion Stroke

    When the cylinder's contents are compressed as much as possible, the spark plug ignites the gasoline and air within the cylinder; this causes the cylinder to explode, driving the piston down.

4. Exhaust Stroke

    Once the piston hits the bottom, an exhaust line opens, allowing the exhaust to leave the cylinder and travel through the tapeline.

The piston is connected to a crankshaft by a connecting rod. As the crankshaft revolves, it resets the cycle consistently; this allows the four-stroke combustion cycle to continually repeat itself, ensuring that the car doesn't stop moving until it is out of fuel.

JET ENGINE

A jet engine uses the same principle as an internal combustion engine: burning gas with air to release energy. However, instead of cylinders, like the internal combustion engine, the jet engine uses a long metal tube. All four steps that occur within four separate cylinders in the internal combustion engine all happen in a straight-line sequence within the tube. Three clear differences make a jet engine much stronger than a car engine:

1. A jet engine is devised to take in much more fuel than a car engine, resulting in the force and energy being generated to be vastly superior to that of a car engine.
2. Due to the lack of cylinders, all of the steps within a four-stroke internal combustion engine can occur simultaneously; this allows the jet engine to output the maximum amount of power consistently.
3. The jet engine also makes its exhaust pass through multiple turbine stages, allowing for the extraction of as much energy as possible.

SOLAR CELL

Solar panels have become more popular over the past couple of years due to their ability to produce "green" energy. Interestingly, solar panels consist of many smaller units or solar cells. These cells link up together to create a fully-functioning solar panel.

Each cell is a sandwich made up of two thin slices of silicone. This silicon is "doped" with other materials, giving each slice of silicone a positive or negative charge. When opposite charges are generated, an electric field is created.

This electric field consists of a large number of atoms with valence electrons. When a solar panel is affected by sunlight, the photons within the light knock these electrons out of their atoms, generating electric flow. These electrons are transferred through wires to create a constant electric current.

RADAR

Radar is a large part of the way that our society functions. From air traffic control to NASA, radar is used quite extensively in our society today. When using radar, scientists wish to do three major things:

1. Detect the presence of an object at a distance.
2. Detect the speed and direction of the object.
3. Map out the location of the object.

These three things are all accomplished through the use of echo and doppler shift. Though radar works using radio waves, the actual principle can be explained better with sound waves.

An echo is a widespread phenomenon that occurs when sound waves reflect off surfaces. When you yell into a canyon or cave, your sound is heard a bit later; this is because the sound wave is reflected back to you after hitting the canyon or cave walls. The time taken for the sound to return to your ears depends on the distance the sound wave has to travel to get back. So the longer the length, the more time it takes for the sound to return to you.

Like the Doppler Effect explained previously, the Doppler Shift occurs when a sound is propagated from a moving object. The amount of time for the sound wave to travel is decreased. For example, if there is a car a bit of distance away and it toots its horn. You will hear some delay before hearing the sound depending on how long it takes for the sound wave to travel that distance. Yet, if the car is moving towards you and blows its horn, you will still hear some delay, but the actual sound will last for a shorter period of time. The reason for this is by the time the sound of the horn reaches you; the car would be right next to you.

When you combine an echo and the doppler shift, you get the basis of a radar. Some sound waves will bounce off the car; note that these reflected sound waves will be compressed since the vehicle is moving. So, the sound has a higher pitch than the original; by measuring the echo's pitch, one can determine how fast the car is going.

Radar, specifically used to navigate airplanes, uses its transmitter and shoots out high-intensity, high-frequency radio waves. After this, the radar turns off its transmitter, turns on its receiver, and listens to the echo. In ground-based radar, there is a lot more potential for interference. For example, in police radar, a wave can bounce off various objects before returning to its source, leading to the necessity to filter out this clutter.

INTERESTING DISCOVERIES

Overall, several different aspects of physics remain mysterious to humans. Today there have been various phenomena witnessed, which still have no concise present-day explanation. The ones that will be discussed in this book are sonoluminescence and sonofusion.

SONOLUMINESENCE

Sonoluminescence is a mysterious phenomenon whereby sound inside a bubble is turned into heat and light. In order to achieve this occurrence, a small flask of water is surrounded by two ultrasound piezoelectric speakers tuned to resonate the water-filled flask's frequency; this varies based on the flask's material and shape. When the sound wave's pressure is greater than atmospheric pressure, bubbles of dissolved gas form due to the negative pressure. As the pressure in the flask becomes positive, the bubble collapses and then oscillates before expanding again. As the bubble collapses, the air inside heats up, resulting in a flash of light. There is heavy debate among scientists about whether this phenomenon can be used for nuclear fusion.

SONOFUSION

Thermonuclear fusion has consistently held the promise of cheap, clean, and limitless energy. Yet, creating a reactor that produces more energy than it consumes has not yet been identified. Nonetheless, scientists have recently found new mechanisms for inducing fusion, using sonoluminescence. As stated previously, if the same experiment of sonoluminescence is conducted with a flask filled with a deuterium-rich liquid, when the bubbles filled with deuterium vapor collapse, it could result in some of the deuterium nuclei undergoing fusion. For reference, deuterium is also known as heavy hydrogen, an isotope of hydrogen consisting of one proton and neutron. Thereby giving it double the mass of the nuclei of an ordinary hydrogen atom.

Since discovering sonoluminescence, researchers have speculated that the temperature and pressure inside the bubbles could reach fusion-level necessities. To achieve these circumstances, scientists have tried to use more enhanced forms of sonoluminescence, known as single-bubble sonoluminescence (not named as creatively as the idea itself). In the most recent version of this experimentation, after the formation of bubbles in the flask, scientists fire a neutron that goes through the liquid. These neutrons collide with the molecules in the liquid resulting in tiny clusters of about 1000 bubbles. These bubbles instantly swell (a process known as cavitation), then the pressure cycle rapidly reverses, and the bubbles implode with a huge amount of force.

This implosion creates shock waves within the bubble that travel at high speeds that converge at the bubble's center. This shock wave creates an extreme amount of pressure, which causes the deuterium nuclei of the liquid to collide at incredible speeds. This pressure is enough to result in fusion. The main difference between this method and normal sonoluminescence is the increased pressure, the firing of neutrons, and the liquid-rich in deuterium.

Now that the circumstances for nuclear fusion are created harnessing the energy from these reactions is a separate challenge. This whole reaction only exists for a very brief amount of time (about a picosecond or one trillionth of a sec or 10^{-9} sec) and is confined to a tiny region. The output energy is minimal, hence why the whole cycle doesn't meltdown the container where they occur. In order to obtain a steady flow of energy, the next step would be to scale up the apparatus and make the fusion self-sustaining, a point that scientists still have not achieved.

THEORY OF RELATIVITY

Albert Einstein in 1905 described the special theory of relativity. Later this was enhanced further, and in 1916 the general theory of relativity was formulated.

SPECIAL THEORY OF RELATIVITY

Suppose you are traveling in an airplane. Only an observer outside the aircraft can determine if the plane is moving or not. The individual inside will be unable to tell without looking outside. Hence all the Newtonian concepts of displacement, velocity, speed, acceleration, etc., are valid with respect to a frame of reference. In the airplane example, the observer outside the airplane is the frame of reference necessary to determine the plane's movement; this is true for stationary objects on Earth. If you see a rock on a mountain to be stationary, it is since the observer, i.e., you are on Earth. As you know, Earth is moving around the Sun, and if the observer is in space, then he will not find that rock as stationary.

These frames of reference fall apart when we try to calculate the speed of light with respect to an observer. It is known that the speed of light 'c' is constant in a vacuum, equal to 299,792,458 m/s. This value of the speed of light in the vacuum always stays constant irrespective of whether the observer is moving or not. Even if the light source moves at the speed of light, we always determine this value to be equivalent to 'c.' It means the speed of light neither depends on the speed of the source nor the observer's speed.

The special theory of relativity has two primary principles. Firstly, all the laws of motion described by Newton and others are valid only with respect to an internal frame of reference. And secondly, the speed of light is always constant to any frame of reference. The second principle contradicts the first principle when "observing" the speed of light. This inconsistency raises many questions about our concepts of space and time.

There are two consequences of the special theory of relativity: time dilation and length contraction.

TIME DILATION

Suppose a spacecraft is traveling near the speed of light. All the people traveling in that spacecraft will observe that the motion of all bodies is following the two principles of the special theory of relativity stated above. According to Newton, even their bodies, heartbeat, etc., will follow all the laws of motion. Now, imagine a person on Earth, at rest, compared to the spacecraft traveling almost at the speed of light. A healthy person living on Earth will live for around 100 years. However, how long would the people traveling in the aircraft live? Hypothetically, this person on Earth has a twin who is located on the spacecraft. What would be the twin's age in comparison to his sibling on Earth?

Note that time normally moves in their unique environments for both the twins. Only when the spacecraft returns to Earth we will notice a huge difference. The person living on Earth will be much older than the twin who traveled on the spacecraft; this is called time dilation, and it is expressed in the following famous formula.

$$t^{sc} = t^{E} \sqrt{1 - V^2/c^2}$$

Where: t^{sc} = Dilated time on the spacecraft
t^{E} = Time on Earth or stationary time
V = Velocity of the spacecraft wrt Earth
c = Speed of Light

FIGURE 96: EQUATION FOR TIME DILATION

Using the above formula (which was proven in experiments conducted at the CERN laboratory in Geneva using fast-moving atomic particles (muons), we can measure the time taken for the spacecraft to travel with respect to Earth. This formula shows that clocks on moving objects and stationary objects will differ based on their velocity. As the speed of light is extremely high, in comparison to us, we do not notice the time dilation between moving and stationary objects on Earth. In the case of a spacecraft which travels near the speed of light, we will witness time dilation.

LENGTH CONTRACTION

Assuming the spacecraft in the previous example is traveling to a distant planet, if an observer on Earth were to calculate the distance to the planet, their value would be much larger than if an observer from the spaceship were making the same calculations. For the person traveling in the spacecraft, the distance to the planet will be much smaller than the distance calculated on Earth. As we know, distance is speed multiplied by time; hence if time differs because of time dilation, then length also changes for the two observers.

Another consequence of the special theory of relativity is that all length and distance contract (which include the distance traveled, the length of the spacecraft, etc.) as we move closer to the speed of light. By that definition, space contracts as well. This space contraction will not be felt by the people traveling in the spacecraft. Observers outside the spacecraft will notice. If the spaceship is moving in the horizontal direction, then the length contraction happens only horizontally. The length in the vertical direction remains intact; this is due to length contraction only applying in the direction motion of the spacecraft.

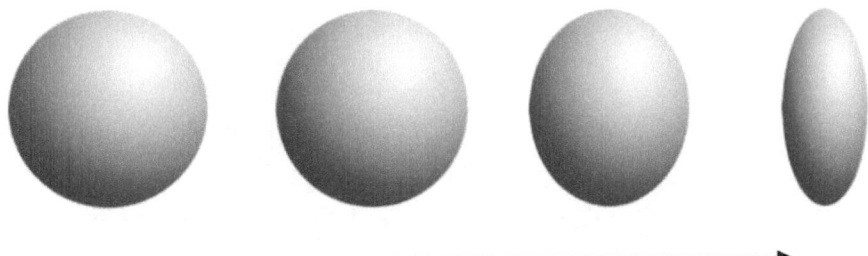

Planet will length contract in the direction of motion of the spacecraft as the spacraft travels near the speed of light

FIGURE 97: PLANET'S LENGTH OR DIAMETER CONTRACTION

The formula for length contraction below can be derived easily from the time dilation equation.

$$l^E = l^{SC}\sqrt{1 - \frac{v^2}{c^2}}$$

l^E = Length of the spacecraft measured from Earth

l^{SC} = Length of the spacecraft measured in the craft

V = Velocity of the spacecraft wrt Earth

C = Speed of light

FIGURE 98: LENGTH CONTRACTION FORMULA

GENERAL THEORY OF RELATIVITY

In the special theory of relativity, we never considered the effect of gravity. It is known that all objects are subjected to gravitational attraction. Gravity creates another kind of mass calculation for all the objects. The first one we learned earlier in Kinematics is the inertial mass, and the second one is gravitational mass; this is produced due to gravitational attraction. It has been proven in several experiments that both gravitational mass and inertial mass are equal. Still, this gives rise to a question as to why these two different types of mass (inertial and gravitational mass) should be equal? And why gravity should be related to a body's inertia?

To answer this question in 1916, Einstein published his theory of general relativity, which determined that massive objects distort space-time, which is felt as gravity. This idea might sound complicated at first, but it can be divided into a couple of simple key concepts.

GRAVITY

Sir Isaac Newton quantified gravity when he determined his three laws of motion. According to Newton's Law of Gravitation, the gravitational attractive force between two objects depends on how massive and how far apart the bodies are. While Newton considered gravity to be a force that acted on a distance, Einstein proved that the laws of physics are consistent for all non-accelerating observers, where the speed of light in a vacuum is fixed, no matter the speed of the observer. Therefore, space and time are woven into a single continuum known as space-time. So, unlike what was previously believed, space and time work as one entity, where the manipulation of one could manipulate the other as well. So, events that occur simultaneously for one observer could occur at different times for another. By introducing the concept of space-time, the acceleration due to gravity becomes relatively easy to explain. In short, Einstein postulated that we do not need a gravitational attractive force to account for gravity. Instead, the space-time curvature near massive bodies is enough to explain the reason for gravity.

DISTORTION IN SPACE-TIME

Massive objects, due to their large mass, result in a distortion in space-time. This idea is commonly shown with the trampoline and bowling ball "experiment." Imagine a large trampoline that represents space-time on a single plain. And now, place a large bowling ball (an incredibly massive object) in the middle of this trampoline. The bowling ball would press down into the trampoline resulting in a dimple. If a marble (a less massive object) is rolled around the edge, it will spiral towards the bowling ball, the same way the gravitational pull of large bodies affects smaller bodies. This spiraling effect is what we notice as acceleration due to gravity. Note, this distortion of space-time happens in all three dimensions and is difficult to visualize on a two-dimensional piece of paper. The following is the best depiction of space-time distortion.

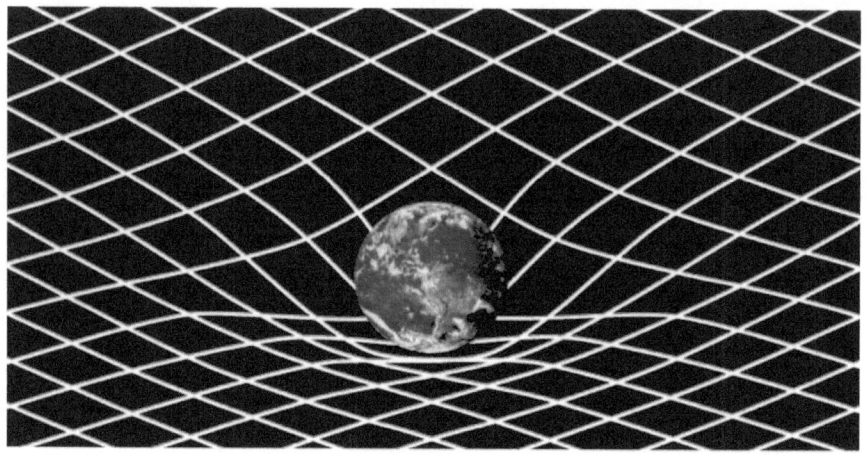

FIGURE 99: EARTH'S MASS DISTORTING SPACE-TIME [34]

All objects falling to Earth follow this spiral path; yet, we see them falling in a straight line due to these objects being observed from Earth. This spiraling effect can be observed with the satellites revolving around the Earth. In fact, this is the reason the satellites do not fall straight to Earth. Satellites traverse in a spiral path, and if the distance is far, they start orbiting around the Earth and slowly come closer. In effect, space itself is curved. Hence all the objects follow this curved path, including light. Objects moving in this space-time continuum follow a natural path, and we see that as the falling of an apple from a tree, orbits of planets, satellites, etc. This method of thinking has led to space being thought of as a sponge that warps depending on the circumstances.

As we learned previously, the concept of gravity and motion influences time. For objects moving faster, their time moves slower. It is also known a gravitational field causes gravitational redshift and gravitational blueshift. Hence, it is crucial to visualize the emptiness of space as a space-time continuum. In effect, we see curved space-time near all massive objects like planets, stars, galaxies, etc.

GRAVITATIONAL LENSING

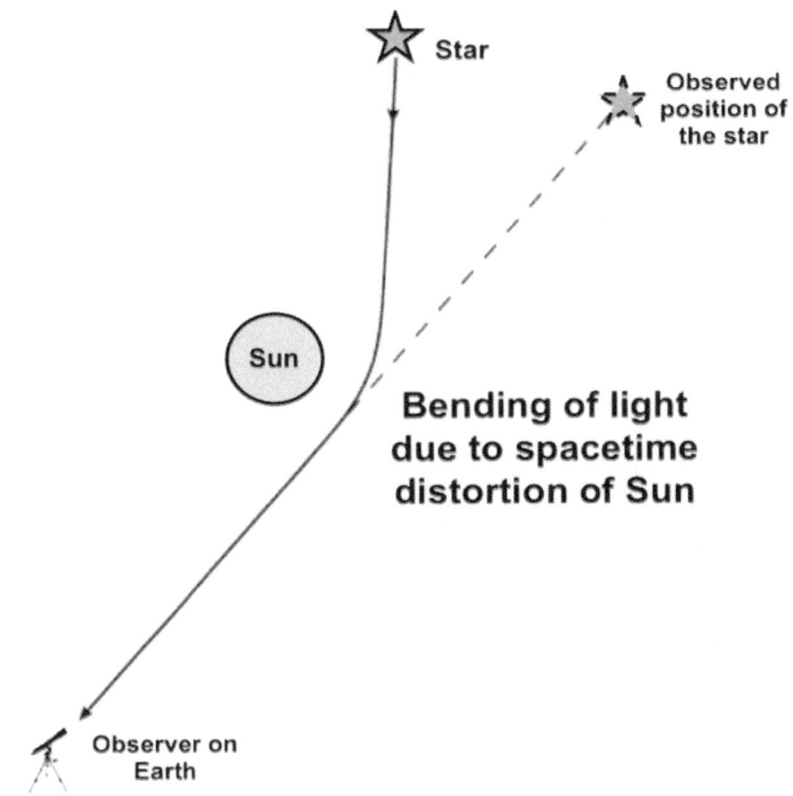

FIGURE 100: BENDING OF LIGHT DUE TO SUN'S MASS

The light around incredibly massive objects bends due to a distortion in space-time caused by the object. This bending of light allows it to be used as a lens for astronomers to study stars and galaxies behind massive objects. Gravitational lensing is also why it is difficult to find and see blackholes; due to their massive gravitational pulls, black holes bend light around them. So, the result is black holes appearing to be invisible to the naked eye.

QUANTUM THEORY

Richard Feynman once said, "I think I can safely say that no one understands Quantum mechanics." In many respects, what Feynman said is true as probabilities play a major role in our understanding of quantum theory. Subatomic particles (quarks, gluons, etc.) play many games with our observations, and hence it is difficult to state their behavior deterministically. Their behaviors can be stated and understood only with a certain amount of probability. Sometimes because of the high degree of uncertainty and paradoxes in quantum mechanics, it sounds like a metaphysical or spiritual subject.

Quantum mechanics is a focus of physics related to the incredibly small that emerged in the 1900s. At the scale of atoms and electrons, the previously discussed ideas of classical physics begin to falter and cease to be useful. In classical mechanics, an object exists at a specified point at a specified time. On the quantum level, objects exist in a haze of probability. There is no finite place where they are. This makes the necessity of new laws and new equations to be created, hence the creation of quantum mechanics.

The first problem with classical physics started with the behavior of light. Light made it extremely difficult to clearly state whether it's made up of particles (photons) or waves. In some experiments, light has behaved like a wave (Young's double-slit experiment). In others, it has acted like a stream of particles (photoelectric effect, i.e., emission of electrons when light or photons hit the metal). This dual behavior of light made scientists, for the first time, think of light as a particle and a wave, and this theory continues even today.

BLACKBODY RADIATION

It is assumed in classical physics that a black body (a body that absorbs perfectly all the radiation hitting it and remits all the absorbed radiation) when heated, the temperature would increase consistently over time. Note there are many materials in nature that behave almost like theoretical black bodies. Yet, upon a series of experiments conducted in the 1900s by German physicist Max Planck, he found that objects heat in steps or "quanta" or packets of energy. He proposed that the energy of electromagnetic waves could be quantized rather than a continuous curve. Additionally, the energy of each of these packets or quanta is proportional to the frequency of the radiation; this means for each temperature, there is a maximum amount of energy emitted by the black body corresponding to its energy level.

Therefore, energy can be lost or gained, not as a continuous function, but in terms of small step units. And as the frequency or energy in an electromagnetic wave increases, the magnitude of the quantum or radiant energy also increases. Neil Bohr proposed (based on the results of Rutherford's double-slit alpha particle experiment) the structure of an atom in which electrons revolved around the nucleus in different energy levels (1, 2, 3, etc.) or orbits. Also, Bohr proposed as long as the electron stayed in a given energy level, it does not lose or emit energy. For this reason, when an electron drops from a high energy level to a lower one, it emits light. The energy level that the electron drops is a quantum. Thereby making every energy transfer in the world based on a quantum scale instead of a linear one. Bohr's atomic model explains the behavior of black body radiations and quanta and many other experimental results such as the Crompton scattering of X-rays.

DOUBLE-SLIT EXPERIMENTS

Later on, de Broglie's theories and experiments of George Thomson proved electrons also sometimes behaved like waves. The double-slit experiment with a stream of electrons again proves the wave properties of electrons (we view an interference pattern on the screen behind the double-slit, like waves when a stream of electrons are fired through it). In turn, these discoveries prove that even atomic particles like electrons behave like waves and also as particles. In classical physics, we say the electron must travel through the first slit or the second, but in the quantum world, we say the electron could have traveled through the first slit, second, or both since it behaves like a wave.

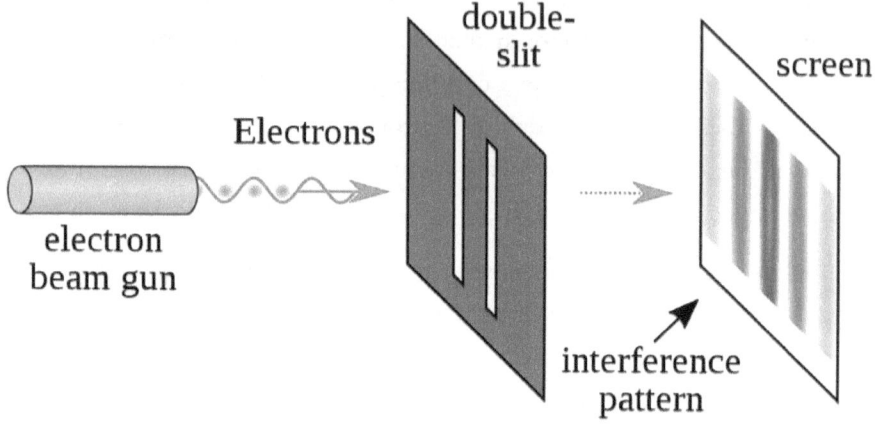

FIGURE 101: DOUBLE-SLIT EXPERIMENT WITH ELECTRONS [35]

Also, note Thomas Young proved light behaved like a wave using the double-slit experiment. Neil's Bohr used the results of Rutherford's double-slit experiment with alpha-particles to propose the structure of an atom. This double-slit experiment is exceptionally significant in quantum physics, but no one understands how the particle to wave transition happens in the experiment even today. In the case of electrons (or photons), it is impossible to determine if the electron passed through the first slit or the second. There is always a 50% probability for each slit; this is why quantum mechanics is always probabilistic, unlike classical physics. This concept is called the superposition principle of quantum mechanics, which defines quantum states for a particle with specific probabilities. In short, we can't determine that a particle is either here or there. Instead, we can say a piece of it is here, and another portion is there. Therefore, we can't find the particle's position with certainty, which is formulated in Heisenberg's uncertainty principle.

HEISENBERG'S UNCERTAINTY PRINCIPLE

Simply put, the Heisenberg Uncertainty Principle states that there is no way to know the exact speed and exact position of a particle. In quantum mechanics, the exact position and exact speed of a particle have no meaning since quantum particles behave as both waves and particles.

To further understand this, we need to know what it means to act as both a particle and a wave. A particle has a definite position over time. There is a 100% probability that a particle is in one position and a 0% probability that a particle exists in another position simultaneously. In other words, a particle cannot exist in two places at once. On the other hand, waves are spread out throughout a space, such as ripples in a pond. We cannot assign waves a single position because the wave has a probability to exist in multiple positions at once since waves are spread throughout space.

In waves, one can measure the wavelength, which is related to its momentum. So in quantum particles with measurable wavelengths, there is no way to determine the object's position. On the other hand, in particles, one can determine the position; however, it has no wavelength, thereby having no momentum. This concept is more clearly represented in quantum particles whose dual nature is still truly not understood. As a whole, the Heisenberg Uncertainty Principle isn't due to good or bad measuring. It's more associated with the dual nature of all particles.

When measuring atomic particles such as an electron's position and speed or momentum, we need to use light to observe it. As we know, we can't measure any length smaller than the wavelength of light. Also, when the photon in the light hits the electron, it will change the position of the electron—this is a practical limitation of this method of measurement. Even if we use some other way to measure the position and speed simultaneously, we will face additional problems. The issue arises with the behavior of atomic particles stated in the uncertainty principle. Also, note the measuring pairs such as speed and position or position and momentum or wavelength and pitch are related quantities where one measurement affects the other, and the uncertainty principle holds.

SÖDINGER'S WAVE EQUATION

I will try to explain Schrödinger's wave equation without going into partial differential equations. It is difficult to understand the equation without knowing calculus, and we also need to accept some strange behavior similar to that of children's fairytales.

Schrödinger's wave equation is a partial differential equation describing a system's quantum states with time and spatial position. The equation shows the system's quantum states with evolving space and time and predicts that the system's specific properties will be quantized when measured. Some believe this formula can be applied to all atomic particles (since it describes all the quantum world's strange behaviors concisely) and even the entire universe. For example, an electron's energy and its angular momentum are quantized. This equation also predicts a system can have many quantized energy states, quantum tunneling (wave function or the particle passes through barriers), particles as waves, and measurement uncertainty.

SÖDINGER'S CAT AND SUPERPOSITION

Schrödinger's cat is a famous thought experiment created by Austrian physicist Edward Schrödinger in 1935 to help explain odd behaviors witnessed by Albert Einstein and other quantum physicists. Imagine a cat placed inside a sealed box with a radioactive substance with a 50% chance of killing the cat in an hour. After an hour has passed, what state is the cat in?

The cat is either dead or alive; simple common sense tells us this. Yet, according to quantum mechanics, Schrödinger pointed out that the cat is both dead and alive the moment before the box is opened. We can only determine the definite state of the cat after the box is opened. Up till that point, the cat is an obscure probability.

FIGURE 102: SCHRODINGER'S CAT MAY BE DEAD OR ALIVE

Like Schrödinger's cat, all quantum particles are blurs of probability until they are measured. This phenomenon is known as superposition. In superposition, a quantum particle, such as an electron, can exist in two states at once: a wave and a particle. One of the most famous experiments of superposition consisted of firing photons (light particles) at two parallel slits in a wall (or double-slit explained earlier). The fundamental idea of quantum mechanics is that the photon acts as a wave, thereby interfering with the other photons, creating an iconic wave pattern on the other side of the wall. The odd thing is that the interference also occurs if only one particle is fired at a time. The particle seems to pass through both slits simultaneously, interfering with itself; this means that the photon acts like both a wave and a particle, along with being in two different positions at once.

It only gets weirder: when scientists determine which slit the particle goes through, one invariably finds that the particle only went through one slit, and the wave interference disappears. Leading to the interesting idea that when one is looking, a particle acts like a particle, but the moment one looks away, the particle is now a wave. This marvel is why quantum mechanics is so diabolical for physicists, because the moment one starts to measure, the phenomenon disappears.

QUANTUM ENTANGLEMENT

Many quantum effects can be understood from the wave equation. One of the interesting ones is quantum entanglement. The best way to conceptualize quantum entanglement would be through a simple thought experiment. Imagine two cakes; they can either be square shaped or circular. There are four possible joint states for the two cakes: (square, square), (circle, circle), (square, circle), or (circle, square). The shape of the two cakes in the classical world is independent. This means knowing the shape of one doesn't indicate the other's shape, and vice versa.

Now imagine that the two cakes are entangled. In an extreme form of quantum entanglement, the shape of one cake ultimately indicates the other cake's shape. Therefore if one cake is a circle, the other is also a circle. If one is a square, the other is also a square. So, under entanglement, the two cakes cannot be different shapes.

In quantum systems, entanglement arises naturally (e.g., the aftermath of particle collisions). On the quantum scale, the interactions between particles result in correlations among them—for example, molecules. A molecule is a system made up of electrons and nuclei, and at the lowest energy state, this system is highly entangled. As the nuclei move, the electrons move as well, making their positions, in no way, independent of one another.

QUANTUM COMPUTERS

Standard computers use bits to model specific functions of a computer. Bits are small switches that are either on or off, thereby reflecting a one or a zero. Everything in a computer, from a website to a photograph, is represented by a set of ones and zeros. However, these don't represent the way the universe functions. Everything in the universe is uncertain, not either a one or a zero. Even supercomputers are unable to model uncertainties to the degree necessary. To compensate, scientists created quantum computers.

Instead of using bits, quantum computers use qubits or quantum bits. Qubits usually are subatomic particles such as electrons or photons. In order to manipulate qubits in a quantum state, superconductors are cooled to freezing temperatures, or individual atoms are trapped in electromagnetic fields on silicon chips in ultra-high-vacuum chambers. The goal in both cases is to isolate the qubits in a controlled quantum state.

Unlike bits, qubits are in a state of superposition. As stated before, the idea of superposition is acting as both a wave and a particle at once. While bits are either on or off, a one or a zero, qubits are somewhere in the spectrum between one and zero. Imagine a coin, signifying a bit, in a computer; one would flip the coin, with there being a chance of landing on heads or tails. In a quantum computer, one can spin the coin, signifying a qubit. In this circumstance, the coin is a blur of probability where it's either heads or tails until measured. This position of the coin being in motion can be considered as superposition.

In superposition, several qubits can crunch through a vast number of potential outcomes simultaneously. To illustrate, if a programmer were to tell a regular computer to solve a maze, the computer would travel through the maze using multiple different paths one at a time until it found the correct path. A quantum computer would travel through all the paths in the maze simultaneously due to qubits being in a state of superposition. However, this phenomenon is also a double-edged sword. Though superposition can factor in uncertainty, ensuring a lightning-fast processing speed, the final result of such a calculation only emerges once the system is measured. But, measuring the system results in the quantum state to collapse, reverting to a binary system.

Researchers can also generate entangled qubits. As explained previously, quantum entanglement between two objects ensures that the outcome of one system is a reflection of the outcome of another. Therefore, when they are entangled in terms of qubits, changing the quantum state of one qubit results in a parallel change of the other qubit. Entanglement in quantum computing is incredibly useful. In an ordinary computer, doubling the number of bits results in doubling the amount of processing power, making it a somewhat linear relationship. In a quantum computer, doubling the number of entangled qubits results in an exponential increase in the processing speed, allowing these computers to crunch numbers at astronomical rates.

This may sound amazing, and one might be wondering why there isn't a quantum computer located in every home. That is due to decoherence. The quantum state is extremely fragile. Even the slightest alteration of the environment, or "noise", can result in qubits tumbling out of superposition before the task is done. For this reason, researchers try to protect qubits by keeping them in vacuum chambers and supercooled fridges. Despite their best efforts, still, noise can cause massive amounts of the error to creep into calculations. Smart quantum algorithms and the addition of qubits have been used to decrease error, yet, in order to achieve the reliability necessary, thousands of qubits need to be added. This vast increase in qubits will sap the computer's computational capacity. Additionally, scientists can't generate more than 128 qubits.

Quantum computers are quite far away from achieving their full potential. Still, scientists are eager to see the immense benefits that they can bring to the world. Ranging from finding vaccines to calculating optimal routes for buses and trains, quantum computers would, theoretically, be able to do that and more.

PHOTONS

The nature of light has always been a heavily debated topic among physicists until it was resolved approximately a century ago. Albert Einstein introduced the duality of light as being a particle, or photon, where the stream of photons is a wave. The reason for light's particle nature was the photoelectric effect, in which electrons flow out of a metal surface exposed to light. This would not have been possible unless the light is made of quantum particles.

Einstein speculated that during the photoelectric effect, electrons within the metal collide with photons. These electrons take the energy from the photons, resulting in them flying out. The overall concept of Einstein's theory was that light's energy is related to its oscillation frequency; this would mean that the intensity of light is reflected by the number of photons in the beam.

Since photons are light particles, they can travel in a vacuum at the speed of light and are, in fact, always in motion at the speed of light. These particles have no mass and no electric charge. Understanding a photon also reflects the importance of the speed of light in Einstein's theory of relativity. The photon and its interactions with space and time are reflected when Einstein showed the relationship between mass is equivalent to energy. Though there is more research being done on the photon's logistics, in general, scientists see the universe in a new light due to this discovery.

MATTER AND PARTICLE PHYSICS

Particle physics is an aspect of physics that focuses on the incredibly small particles that make up all matter and radiation and their interactions. It is also known as "high energy physics" since these elementary particles don't occur naturally but are created and detected during high energy collisions with other particles. Such collisions are commonly done in particle accelerators.

Modern-day particle physics mainly focuses on the subatomic particles that structure atoms. These include electrons, protons, and neutrons, with protons and neutrons being made up of quarks. In particle physics, the use of the term 'particle' is a bit of a misconception. Usually, the particles studied aren't necessarily particles but more waves since they are governed by quantum mechanics. These particles are traditionally studied in experimental conditions where they show wave-like properties.

All the particles and their interactions with one another through observed data are described through the Standard Model. The Standard Model contains 18 "species" of elementary particles that can combine to form other particles. This model continues to grow as scientists run experiments and determine results.

ATOMS

The general theory of the atom, small particles making up every object in the universe, has been dated as far back as 440 B.C. And the atomic model has consistently been added and modified for the past 200 years to determine an atom's exact make-up. After constant experimentation, scientists decided atoms to be the basic units of all matter.

Atoms bond with other atoms through the sharing or taking of electrons. When atoms bond, molecules are created. For example, the water molecule (H_2O) is one oxygen atom bonded with two other hydrogen atoms. Billions of molecules make up a simple glass of water. These numbers are so large that chemists use a particular scale based on moles to measure them (as previously discussed).

The structure of an atom consists of a central nucleus, which contains protons and neutrons. Since protons have a positive charge and neutrons have no charge, this makes the nucleus, overall, positively charged. Similar to how the Sun has planets orbiting it, the nucleus has electrons orbiting it. The atom's overall mass is concentrated at its nucleus since protons and neutrons have equal mass, but the mass of an electron is almost nothing. Since electrons are negatively charged, and the neutrons are positively charged, they attract. The force that keeps neutrons and protons together is known as the strong force. The strong force holds together the atomic nucleus and is the basis for all interactions containing quarks (we will discuss this further later).

Electrons orbit the nucleus in shells at varying energy levels. The farther the shells are from the nucleus, the more energy they have. Additionally, the substance's identity is determined by the number of neutrons and protons an atom has. Depending on how this number changes, the object can also change. Scientists run experiments in particle accelerators where atoms are bombarded with protons to change their atomic numbers and properties.

The term 'atom' originates from the Greek word for indivisible because, initially, atoms were thought to be the smallest things in the universe and couldn't be divided any further. However, after much technological development, scientists have determined that an atomic nucleus can be further divided into quarks and other particles. Now let's focus on the make-up of an atom.

ELECTRONS

Electrons are subatomic particles known to orbit the nucleus of an atom (as stated previously). Electrons can behave as both a wave and a particle. As proven by the Heisenberg Uncertainty Principle, it is impossible to know the exact location and speed of an electron. For this reason, most atomic models have an electron cloud surrounding the nucleus of the atom instead of the defined circular orbits that one usually uses to describe an atom.

Electrons are integral when it comes to the bonding of different atoms. The interaction between the outer electron layers of separate atoms is known as bonding. Bonding can occur in two forms, either covalent or ionic. Ionic bonding is where an atom gives up electrons to another atom entirely. Covalent, on the other hand, is when atoms share electrons. Every atom aims to gain 8 electrons (the actual capacity of electrons in an Nth orbit is $2N^2$) in its outer rings. This ensures that they are 'complete.' Bonding is a means that atoms use to help gain these electrons, specifically for atoms that don't originally contain many electrons in their outer shell. It is for this reason specific substances are incredibly reactive with others. For example, potassium is incredibly reactive with fluorine. Potassium has one electron in its outer shell, and fluorine has seven, thereby making it a match made in heaven for an ionic bond.

Ionic bonds are stronger than covalent bonds. Since one atom takes the other electrons, there is a strong, attractive force between them, stronger than that of a covalent bond. Additionally, each electron orbit can only hold so many electrons. Atoms can only bond if there is room to share and receive extra electrons. This is why the noble gases are considered incredibly unreactive, by this same logic because they already have 8 electrons in their outer shell, thereby making bonding unnecessary.

Unlike protons and neutrons, electrons are not made of quarks. As of right now, scientists are unable to determine whether an electron can be split into smaller pieces. Therefore electrons are known as fundamental particles, similar to quarks since they cannot be built out of anything smaller.

NUCLEUS

As stated previously, the nucleus is the atom's center and consists of both protons and neutrons. A neutral atom contains an equal amount of protons and electrons, but the number of neutrons in a particular element can vary. Atoms of the same element that have varying numbers of neutrons are called isotopes. Additionally, neutrons are incredibly microscopic and also massively dense in comparison to the rest of the atom. Now let's discuss the makeup of the nucleus; protons and neutrons.

PROTONS

The discovery of the proton was first witnessed by JJ Thomson in 1910 and Willhelm Wien in 1898 when studying streams of ionized gaseous atoms and molecules whose electrons had been stripped. Thompson and Wien identified a positive particle equal in mass to the hydrogen atom. Rutherford furthered this concept in 1919 by showing that nitrogen ejects "hydrogen nuclei under alpha-particle bombardment." By 1920, this "hydrogen nucleus" was an elementary particle named the proton.

In the late 20th century, the study of high-energy particle physics refined the structural understanding of the proton. It was found that protons are made up of smaller particles, classified as baryons, known as quarks. Additionally, protons from ionized hydrogen are given high velocities in particle accelerators and are commonly used as projectiles to produce nuclear reactions. Hence, protons are the chief product of some types of artificial nuclear reactions.

NEUTRONS

The existence of a neutron was theorized by Rutherford in 1920 and discovered by Chadwick in 1932. After this finding, scientists clambered to study the properties and interactions of this particle. It was found that when various elements are bombarded with neutrons, they undergo fission. Fission is a type of nuclear reaction in which a heavy element's nucleus is split into two nearly equal smaller pieces. Additionally, each fission-ed nucleus gives off additional free neutrons; enough of these neutrons are produced to sustain a chain reaction. This advancement was used to construct the atomic bomb.

A free neutron is also subject to radioactive decay called beta decay. Unlike alpha decay, a neutron breaks down into a proton, an electron, and an anti-neutrino during beta decay. Due to that fact, this neutron readily disintegrates; it doesn't exist in nature in its free state, except in cosmic rays. Since free neutrons are neutral, they can easily pass through electric fields, thereby being a penetrating form of radiation.

Neutrons, along with protons, are classified as hadrons, subatomic particles subject to the strong force. Like protons, neutrons have an internal structure made up of quarks. The neutron possesses a magnetic dipole. Therefore, it behaves in ways that suggest it is a particle of moving electric charges.

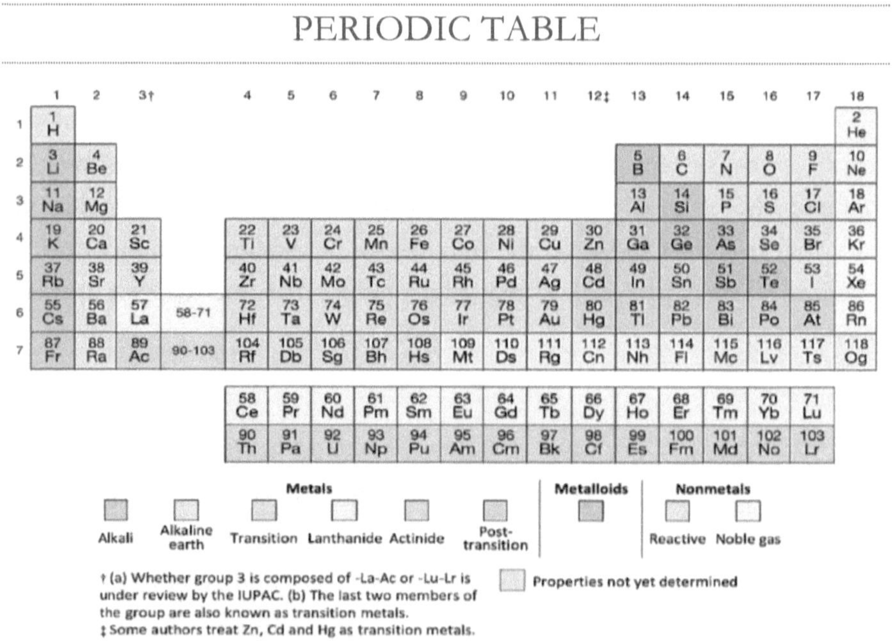

PERIODIC TABLE

FIGURE 103: PERIODIC TABLE OF ELEMENTS [36]

The periodic table of elements arranges all known chemical elements in an informative array. Elements are arranged from left to right as well as top to bottom based on increasing atomic number. The order of these elements coincides with increasing atomic mass. The rows of the periodic table are known as periods. The value of the period for an element signifies the highest energy level of electrons in that element (while the element is not excited). The number of electrons in a period rises as one moves lower the periodic table. Therefore, as an atom's energy level increases, the number of energy sublevels per energy level also increases.

Elements that occupy the same column on the periodic table are called a group. Groups have identical valence electron configurations and behave similarly during chemical bonding. For instance, all the elements in group 18 are noble gases. There are several different sections on the periodic table.

ATOMIC NUMBER

The number of protons in the nucleus of an atom is referred to as the atomic number. The number of protons is what defines the chemical behavior of the element. Note, the number of protons in an atom remains constant; yet, there can be atomic isotopes with more or fewer neutrons.

ATOMIC SYMBOL

The atomic symbol (or element symbol) is an abbreviation appointed to represent an element. These symbols are used internationally to determine the element. For example, the symbol of tungsten is 'W' because another name of the element is Wolfram. Additionally, the atomic symbol for gold is 'Au' because in Latin gold is Aurum.

ATOMIC WEIGHT

The standard atomic weight is the average mass in atomic mass units (amu) of an element. Individual atoms have an integer number of atomic mass. This atomic mass on the periodic table is a decimal because it's the average of various isotopes of an element. The average number of neutrons can be found by subtracting the number of protons (atomic number) from the atomic mass.

For naturally occurring elements, the atomic weight is calculated by averaging the weights of naturally found isotopes of the element. But, elements 93-118 are not naturally occurring; they are lab-created transuranium elements. There is no real "natural" abundance. Therefore the convention changes to list the atomic weight of the longest-lived isotope in the periodic table. This atomic weight should be considered provisional since a new isotope with a longer half-life could be produced in the future.

In this category of superheavy elements (whose atomic numbers above 104). The larger the atom's nucleus due to protons, the more unstable the element becomes (note it is possible that more sable isotopes exist for these heavy atoms but we are not able to create them so far). Therefore, these outsized elements quickly disappear, lasting mere milliseconds before decaying into their light counterpart elements.

OTHER ATOMIC PARTICLES

New atomic particles are continuously being discovered as more and more experiments are being done. Hence the creation of the Standard Model to organize these ongoing atomic particles. There are many such particles, thus why we will be going through the most "commonly known" or most experimentally witnessed particles: neutrinos, quarks, and Higgs Boson.

NEUTRINOS

A neutrino is a subatomic particle quite similar to an electron but quite smaller in mass and with no electrical charge. Neutrinos are considered one of the most abundant particles in the universe because they have very little interaction with matter. Therefore, they are incredibly difficult to detect, thereby remaining largely a mystery to physicists.

In the late 19th century, researchers were puzzled over the concept known as beta decay, where the nucleus inside an atom emits an electron. This phenomenon seems to violate two fundamental laws of physics: the conservation of energy and the law of conservation of momentum. In 1930, physicist Wolfgang Pauli proposed that an extra particle, known as a neutrino, might be flying out of the nucleus, carrying the missing energy and momentum. A quarter-century later, physicists Clyde Cown and Federick Reines built a neutrino detector and could snag trillions of neutrinos flying out from an atomic reactor.

Overtime, physicists, eventually found that neutrinos come in three distinct variations. The most basic form of a neutrino is the electron neutrino; still, two other types also exist a muon neutrino and a tau neutrino. The Sun has been known to produce a colossal number of neutrinos that bombard the Earth daily. As these neutrinos pass through the distance between the Earth and the Sun, they oscillate between these three types. Hence, in early experiments, two-thirds of the total number of neutrinos were always missing since only the first one was being addressed.

This theory gave birth to another issue; only particles with mass can undergo oscillation—thereby contradicting the earlier ideas of neutrinos being massless. Though the exact mass of a neutrino is still unknown, experiments have determined that the heaviest of the three is at least 0.0000059 times smaller than an electron's mass.

Since there is so little known about neutrinos and their properties, scientists continue to run tests to determine more about these particles. Recently, researchers detected a new type of neutrino called a sterile neutrino. This finding corroborates an earlier scientific anomaly seen in the Liquid Scintillator Neutrino Detector (LSND). This new discovery doesn't fit the known physical framework of the Standard Model, which completely reorients all of known physics.

QUARKS

In fundamental physics, we are told that everything around us is made up of atoms. Additionally, we are told that atoms are made from protons, electrons, and neutrons. Yet, then the question arises: what are protons, electrons, and neutrons made out of? The answer being quarks.

Quarks are particles that combine to form hadrons, such as protons and neutrons, thereby making them one of the fundamental particles of physics. A fundamental particle of physics means that quarks don't seem to have any structure and are indivisible. Interestingly, quarks are not observed independently but always in combination with other quarks. This property is known as confinement. Confinement makes determining the properties of a single quark somewhat impossible to measure directly.

There are six distinct quarks: up, down, strange, charm, bottom, and top. The variation of the quark determines its properties. The up-down quarks makeup protons and neutrons and are the lightest and the most stable. More massive quarks are produced in high-energy collisions and quickly decay into their more stable forms (up and down quarks). Each quark has a different quality called color (this is not the object's physical color). Color refers to how the strong force holds these particles together; this is carried by gluons.

GLUONS

Gluons are a messenger for the strong force and bind quarks together. At the subatomic level, protons behave like in a New York metro at peak hours. When a metro is congested, if somebody falls, everybody feels it. In the same way, gluons do not behave as single identities but show collective behavior. Finding proof of this dense gluon material would lead to the possibility for a new state of matter, completely revolutionizing physics.

HIGGS BOSON

All fundamental fields are associated with a particle. For example, electric fields are associated with an electron. The Higgs Field, named after physicist Peter Higgs, is considered a universal field throughout our universe that interacts with all objects to give them mass. The interaction between an object and the Higgs Field is what provides this object with its mass. The more interaction there is, the more massive this object is.

For centuries, scientists knew that mass was a basic property of all matter. But, there was no idea from where objects achieved their mass. The Higgs field was a scientific theory that was used to answer this question. Yet, there was no real proof of its existence, until recently. In 2012, the ATLAS and CMS experiments at CERN's Large Hadron Collider announced that they had witnessed a new particle. This particle was dubbed Higgs Boson or "the God Particle" due to its incredibly critical mass.

The mass of the Higgs Boson is 126 billion electron volts, or about 126 times the proton's mass. This exact amount of mass is required to keep the universe on the brink of instability. Though some particles have no interaction with the Higgs Field, thereby giving them no mass (e.g., the photon). So, Higgs Boson is the lynchpin in the Standard Model used to explain and characterize fundamental particles in the universe by scientists. Though this discovery is quite mind-blowing, there is honestly very little genuinely known about the particle. And we are only on the brink of the unlimited possibilities that have been opened.

MOLECULES

Molecules are a group of two more atoms that have formed a unit with one another. In various ways, molecules bond with each other, but it occurs explicitly due to intermolecular forces between the two atoms. There are two main types of molecular bonds: covalent and ionic.

Covalent bonds are associated with the sharing of electrons. Ionic is where one atom takes all of the electrons located in another. The type of bond that is created depends on the location of specific atoms on the periodic table.

INTERESTING MOLECULES

As a whole, specific atoms can bond with others, and these create a variety of objects with interesting properties. We will be discussing one molecule in particular: Buckminsterfullerene.

BUCKMINSTERFULLERENE

In the 1940s, the architect known as Buckminster Fuller popularized a geodesic dome. A geodesic dome is a thin circular structure that is very rigid and can distribute stress throughout the entire structure. When chemists at the University of Sussex discovered a spherical C_{60} molecule, it was named buckminsterfullerene for the domes that Fuller popularized. While Fuller's domes are a lattice of triangles (hence the reason for their strength), buckminsterfullerene contains a lattice of 12 pentagons and 20 hexagons.

The discovery of this new form of carbon came as a shock to most scientists. Carbon has consistently been intensely studied. But, it comes in two basic structures: diamond or graphite. That concept is now obsolete. This new carbon structure is a single element in a spherical cage, similar to a soccer ball. This interesting structure has uncovered a whole family with different multiples of carbon atoms known as fullerenes.

Like other physical forms of carbon, buckminsterfullerene is black. Unlike other carbons, it's soluble in non-oxygenated organic solvents. Some functional groups of buckminsterfullerene have applications as catalysts for specific fractions. The most noteworthy aspect of buckminsterfullerene is its exceptional stability and resistance to radioactivity and chemical corrosion.

Researchers continue to try to unearth their properties. For example, upon plating potassium into buckyballs (another name for buckminsterfullerene), they become superconductors at a temperature of -427°F. That is the highest superconducting temperature for any organic compound. Additionally, if these molecules are fired at a steel wall at 15,000 mph, they bounce back unharmed. This resilience is beyond any particle ever known and can stand enormous pressure. Buckyballs are so resilient they are considered to be far stiffer than diamonds and also far more flexible. This incredibly adaptive substance could be used to open up a whole world of possibilities.

NUCLEAR PROPULSION

Nuclear thermal propulsion is considered to be the next-generation nuclear technology that would be used in space travel. This technology uses the heat from fission reactions to accelerate propellants such as hydrogen to tremendous speeds. These speeds would allow astronomers to reach Mars in just three to four months, half the time of the fastest possible vehicle that uses chemical propulsion.

Additionally, there is also a hope that these technologies could also be developed for energy production. There is a tremendous amount of heat produced with nuclear fission, harnessing that would allow for both planet Earth and other space outposts to be fueled for years.

DARK MATTER

Since the 1920s, astronomers have hypothesized that the universe contains more matter than can be seen by the naked eye. This "invisible matter" is known as dark matter and roughly makes up 80% of the universe's mass. Dark matter is a new type of subatomic particle that doesn't interact via electromagnetism or strong and weak nuclear forces. Though there is a startling amount of evidence supporting its existence, dark matter remains a hypothesis.

The first inklings of dark matter are associated with the discrepancies that occur near the outskirts of galaxies. According to classical physics, Newton's equations predict that the force that makes stars move in a circular path should be equal to the force due to the star's gravity; this is Newton's universal law of gravitation. Near the center of galaxies, this equation holds. However, the farther one ventures from the center, the two sides of the equation don't match up correctly. Though the amount of discrepancy varies from galaxy to galaxy, the observations were universal.

Such a discrepancy suggested that something was going on that existing theories couldn't explain. Either Newton didn't understand how inertia and gravity worked (which seems unlikely, knowing Newton), or something else was at play resulting in some other force to take effect. The theory used to explain this force was that there is a type of matter that doesn't interact with the light yet exerts a gravitational pull.

Dark matter is the preferred theory because it has done a reasonable job of predicting many of the measurements. Yet, some questions remain. Firstly, if dark matter exists, one should directly observe its interactions as it passes through the Earth or in large particle colliders. And yet, both approaches have been unsuccessful. Additionally, no evidence was found explaining how dark matter affects the relationship between the brightness of galaxies and their rotational speeds.

ASTRONOMY

Humans have always gazed toward the heavens, searching to find meaning in both themselves and the universe. This practice resulted in the creation of a branch of physics known as astronomy. Astronomy is the study of the Sun, Moon, stars, planets, comets, and all other cosmic bodies and phenomena. Astronomy is commonly confused with astrology because both have on cosmic bodies and entities. Yet, they are both separate studies.

Astronomy can also be further condensed into astrophysics. Simply speaking, astrophysics involved studying the physics of astronomy and the properties of objects in space. Within astronomy, there are two key fields: observational and theoretical. Observational astronomers focus on the direct study of cosmic bodies through what can be viewed using technology. On the other hand, theoretical astronomers model and analyze how celestial bodies and systems may have evolved. Depending on the specific celestial bodies' focus ranging from planets to galaxies, an astronomer can focus on that particular aspect of astronomy [37] [38].

SOLAR SYSTEM

Our planetary system is located in the Orion arm of the Milky Way galaxy and consists of the Sun and everything in its gravitational pull. This includes all of the terrestrial planets and Earth, plus the gas giants and dozens of moons and millions of asteroids, comets, and meteorites.

Along with all the planets orbiting the Sun, our solar system is also orbiting the center of the Milky Way galaxy. It takes our solar system around 230 million years to complete one orbit around the galactic center while traveling at about 515,000 mph (828,000 kph).

As of right now, according to sources such as NASA, our solar system, and specifically the planet Earth, is the only one known to support life. Though this might change, we're looking for more out in the vast universe every day.

MOON

The regular rhythms of Earth's natural satellite, the Moon, has ensured tides and a relatively stable climate. The most popular theory about the Moon's origin is that a Mars-sized body collided with the Earth about 4.5 billion years ago; this resulted in debris from both Earth and the body to newly form the Moon in a molten state. Over 100 million years, most of the global "magma ocean" crystallized and eventually formed the lunar crust. The early Moon also developed an internal dynamo, which ensured the mechanism for global magnetic fields. Throughout billions of years, the Moon's surface was bombarded with large amounts of comets and meteoroids, leading to the entire Moon being covered by a rubble pile and debris called the lunar regolith.

There are two sides to the Moon; the light areas are known as the highlands. The dark features are called maria. The dark side of the Moon hasn't been seen by astronomers due to the Moon being gravitationally face locked with the Earth. Even today, there is little knowledge of what is located on the other side of the Moon.

SUN

The Sun is the heart of our solar system and is a yellow dwarf star. It is a ball of gas held together by its gravity, which means that there is no way to stand on it legitimately, and also makes up 99.8% of the mass of the solar system as a whole. At its core, the Sun has an intense amount of nuclear fusion, which fuels the star. Its gravity holds the solar system together, ranging from the smallest particles to the largest gaseous planets. Additionally, the Sun generates electrical currents. A magnetic field is carried throughout our solar system by the solar wind. Solar wind is a stream of charged gas blowing outward from the Sun in all directions.

The interactions between the Sun and Earth drive the seasons, ocean currents, weather, climate, etc. Like the Sun, there are billions of stars scattered throughout our Universe. Additionally, like all things in our solar system, the Sun also rotates. The Sun rotates once on its axis every 35 Earth days.

PLANETS

In this section some of the details about the planets in our solar system are described.

MERCURY

Named after the Roman god of travel, Mercury is the smallest planet in our solar system and the closest to the Sun. Yet, even due to its proximity to the Sun, Mercury is not the hottest planet in the Solar System. Mercury can reach temperatures of about 800°F (430°C) in the day and as low as -290°F (-180°C) at night due to Mercury's atmosphere being unable to retain heat. Instead of an atmosphere, Mercury has a thin exosphere composed mostly of oxygen, sodium, hydrogen, helium, and potassium.

ORBIT AND ROTATION

Mercury has a highly eccentric elliptical orbit. Still, it has one of the fastest rotations, where it can speed around the Sun every 88 Earth days. On the other hand, Mercury spins quite slowly on its axis and completes one rotation every 59 Earth days. Unlike most planets, Mercury does not have seasons; this is since the planet is tilted just 2 degrees with respect to the plane, which means it spins perfectly upright.

COMPOSITION

Mercury is also one of the densest planets in the solar system. It has a large metallic core, which makes up about 85% of the total planet. Additionally, it also has a rocky mantel and a solid crust. Mercury's surface is similar to that of the Earth's Moon and is covered with craters. Note, there are also large areas of smooth terrain. These developed overtime after the planet's interior cooled and contracted.

VENUS

Named after the Roman goddess of beauty, Venus is the second planet from the Sun and our closest neighbor.

ATMOSPHERE

Venus is commonly described as a version of Earth where climate change has gone amok. The thick atmosphere traps the Sun's heat, resulting in a consistently higher surface temperature than 880°F (470°C). However, the higher up one travels the atmosphere, the temperature decreases to being similar to Earth, signifying the possibility of life near the top layers.

This higher temperature results in Venus' surface to be entirely covered by volcanic activity. Venus is also completely covered by craters, but they are quite small; this is due to the thick atmosphere being able to burn up most meteorites before they impact.

COMPOSITION

Since Venus is our closest neighbor, its structure is quite similar to that of Earth. It has an iron core with a liquid mantle and a thin crust of rock on the surface. Even due to the large iron core, Venus's magnetic field is much weaker than that of Earth.

EARTH

The Earth, our home planet, is considered to be the only planet in our Solar System, as of now, that contains life. Unlike the other planets, which are named after Greek and Roman gods and goddesses, the Earth is a Germanic word that simply means "the ground."

ORBIT AND ROTATION

It takes 23.9 hours for Earth to complete one full rotation around its axis. In contrast, it takes 365.25 days to complete one full orbit around the Sun. Contrary to popular belief, seasons are caused by the tilt of the Earth's axis on its rotational plane. During a portion of the year, the northern hemisphere is tilted towards the Sun, and after six months, the situation is reversed. Overall, this results in the Sun being higher in the sky, which changes whether the heating is less or more direct, thereby causing seasons.

COMPOSITION

Due to the molten nickel-iron core, Earth ends up developing an incredibly strong magnetic field. As of right now, a compass always points north. A geologic record shows that there have been times in Earth's history where we have seen a magnetic reversal. Such a reversal is improbable to happen for the next thousands of years. Still, when it has occurred, instead of pointing north, compasses will point South.

MARS

Mars was named after the Roman god of war due to its bright red color. Due to oxidation of the surface, Mars is frequently referred to as the "Red Planet."

ORBIT AND ROTATION

Mars, due to its similarities to Earth, has been intensely studied. Six different spacecrafts orbit Mars, and two robotic spacecraft are at work on its surface. One of the main reasons for this is, Mars is one of the closest planets to ours. Like Earth, Mars completes one full rotation in 24.6 hours and completes one orbit around the Sun in 687 Earth days. Mars is also tilted along its axis, thereby giving it seasons. Unlike Earth, these seasons last longer due to the larger orbit around the Sun.

COMPOSITION

As stated previously, the Martian surface is a bright red due to oxidation of the dirt, leading to it rusting. Mars also shows a watery past with river deltas, lake beds, and specific rocks and minerals which could have only formed due to liquid water being present. Yet, the Martian atmosphere cannot maintain liquid water for too long, and the only water that exists presently is water-ice underneath the surface and on the poles.

ATMOSPHERE

The atmosphere of Mars is incredibly thin and consists of carbon dioxide, nitrogen, and argon gases. This sparse atmosphere does not offer much protection from meteoroids. Additionally, the Sun's heat can very easily escape the planet resulting in the temperature on Mars ranging from as high as 70°F (20°C) to as low as -225°F (-153°C).

Also, unlike the Earth, Mars has no magnetic field. Still, there are areas in the Martian crust that are highly magnetized; this might indicate traces of a strong magnetic field that could have existed in the past.

MOONS

Mars also has two moons, Phobos and Deimos, that may be captured asteroids. Since they may be captured asteroids, they have little to no mass to maintain enough gravity to keep them spherical; this leads to their attractive potato shape.

JUPITER

Jupiter is the most massive planet in the Solar System, about 11 times wider than Earth. Jupiter is a gas planet; its composition is similar to that of the Sun, mostly hydrogen and helium. It is unknown whether Jupiter has a solid core. Still, the planet has a strong magnetic field created due to Jupiter's high-speed rotation, which drives electric current from deep in its core.

ATMOSPHERE

Jupiter's atmosphere consists of three distinct cloud layers. The top layer is made of ammonium ice, while the middle layer consists of ammonium hydrosulfide crystals. The final layer is hypothesized to be made of water vapor and ice. Interestingly, like Saturn, Jupiter also has rings; but, they are not as elaborate and are smaller but made up of dust, not ice.

COMPOSITION

Due to Jupiter being made entirely of gas, there is no solid land to help stop the storms on the planet, hence why Jupiter's spots persist for many years. For this reason, the Great Red Spot has remained a powerful storm for about 150 years. `

MOONS

Jupiter has about 53 confirmed moons and 26 potential moons. Due to its intense gravitational field, Jupiter can attract large amounts of asteroids and meteors towards it, resulting in many moons. The four largest moons are Io, Europa, Ganymede, and Callisto, known today as the Galilean satellites.

SATURN

Named after the Roman titan of agriculture, father of Jupiter, Saturn is the farthest planet from the Earth, discovered by the unaided human eye. It has been known since ancient times.

RINGS

Saturn is the 6th planet in our solar system and the second largest. Saturn's most crowning feature is its large icy rings that surround the planet. These rings are thought to be made of pieces of comets, asteroids, or shattered moons, which were all torn apart by Saturn's strong gravitational pull. These pieces range in size from a pebble to that of a large house. Saturn has a total of 7 rings named A-G in the order that they were discovered. All of these rings are relatively close, except for their distance between Ring A and Ring B, which has a gap known as the Cassini Division.

ATMOSPHERE

Saturn's entire atmosphere is made of clouds and storms. Winds in the upper atmosphere are insanely fast, reaching around 1,600 mph. Along with the intense pressure, it is powerful enough to squeeze these gases into a liquid state. Saturn's north pole has a sizeable hexagon-shaped jet stream with winds traveling at 200 mph. In the center of this jet stream is a massive rotating storm.

Saturn has a magnetic field 578 times stronger than Earth; still, it is smaller than that of Jupiter. Like Earth, Saturn also has an aurore that occurs due to electrically charged particles spiral within the planet's magnetic field lines. While on Earth, this phenomenon occurs due to solar wind, on Saturn, these occur due to charged particles that compose Saturn's rings. These particles interfere with the rapid-moving magnetic field, resulting in this aurore.

COMPOSITION

Like Jupiter, Saturn is also a gas planet and consists mainly of helium and hydrogen. Saturn's core is a denser center of metals and surrounded by rocky materials solidified by the intense pressure and heat. This core is enveloped by a layer of liquid hydrogen, like Jupiter, but much smaller. Therefore, Saturn is less dense than water, leading to it, hypothetically, floating in a bathtub.

MOONS

Saturn has 53 confirmed moons and 29 provisional moons awaiting confirmation. Still, these moons have a wide variety of different worlds and environments that could support life. Enceladus and Titan are home to large internal oceans that can support the underwater life of some kind.

URANUS

Uranus was the first planet discovered with the assistance of a telescope. Uranus was first discovered by astronomer William Herschel and was later universally accepted as a planet through the observations of Johann Elert Bode. He was also the one who suggested Uranus' name, which is based on the Greek god of the sky.

ORBIT AND ROTATION

Uranus makes one complete rotation around the Sun every 84 Earth years. Still, its general rotation is must faster, and one day on Uranus lasts about 17 hours. Interestingly, Uranus is the only planet with a 97.7° angle to its rotation. The common theory is that the tilt occurred due to a collision with a large Earth-shaped object in its past. This angle causes Uranus to have extreme seasons for long periods of time.

COMPOSITION

Uranus is one of two ice giants in our solar system. The majority, around 80% of the planet, is covered in a dense icy liquid (water, methane, and ammonium) surrounding a small rocky core. Yet, Uranus doesn't have a surface for a spacecraft to land on; this is because the planet is mostly fluids. Uranus has a slightly smaller diameter than its counterpart Neptune. It is less dense, making it the second least dense planet in our solar system. Also, similar to Saturn, Uranus also has a series of rings around the planet.

ATMOSPHERE

Uranus' atmosphere mostly consists of helium and hydrogen, along with trace amounts of methane and ammonium. The methane in the atmosphere results in Uranus giving off a blue-green hue. As the solar radiation hits the planet, the methane absorbs all of the red color frequency and reflects that green-ish hue. Wind speeds in this area can reach up to 560 mph and blow in the reverse direction of the planet's rotation. As you get closer to the poles, the winds change direction and blow in the same direction as the planet's rotation.

Uranus also has an interestingly shaped magnetic field. For most planets in the solar system, the magnetic field aligns with the planet's rotational axis. On the other hand, Uranus' magnetic field is tilted at a 60° angle. Therefore, the auroras on Uranus are not aligned with its poles. Additionally, Uranus' magnetic field trails behind the planet for several millions of miles into outer space.

MOONS

Uranus has a total of 27 known moons. The composition of Uranus' inner moons is half rock and water ice. The other's composition is not known, but are likely captured asteroids due to Uranus' strong gravitational pull.

NEPTUNE

Neptune is the only planet in the solar system not visible to the naked eye. Neptune is so far away from the Sun that the amount of sunlight the

Earth receives is 900 times brighter than sunlight on Neptune. Due to its distance from the Sun, Neptune was the first planet discovered based on mathematical calculations and was named for the Roman god of the sea.

ORBIT AND ROTATION

Neptune rotates quite quickly, with a day lasting about 16 hours. And Neptune's orbit around the Sun lasts about 165 Earth years. Interestingly, due to Pluto, Neptune is affected by its gravitational field, and Pluto and Neptune's orbits intertwine. Yet, these two planets can never collide because, for every two orbits that Pluto makes, Neptune completes three. This pattern prevents these bodies from closely approaching one another. Similar to Earth, Neptune has a tilt on its axis, allowing it to experience seasons. Due to the length of each orbit, one season lasts approximately 40 years.

COMPOSITION

Neptune, like Uranus, is the second ice giant in our solar system. Neptune has the same composition as Uranus; yet, it is one of the densest planets in the solar system. Due to this intense pressure, scientists theorized that Neptune might have an ocean of boiling hot water, unable to evaporate away due to extreme pressure.

ATMOSPHERE

Also similar to Uranus, Neptune's atmosphere is made of hydrogen and helium. Unlike Uranus, which omits a greenish-blue color, Neptune is a more vivid blue. As of right now, scientists are unsure of the component that results in this intense color.

Neptune is one of the windiest worlds. Its winds are three times stronger than that of Jupiter and whip frozen methane worldwide at around 1,200 mph. Due to this, a large number of storms appear all over Neptune consistently.

In general, Neptune's magnetic field is about 27 times more powerful than that of Earth. The axis of Neptune's magnetic field, like Uranus, is tilted around 47°. This misalignment results in Neptune's magnetic field to undergo a large amount of variation during each rotation.

MOONS

Neptune has 14 moons, with the largest one being Triton. Interestingly, Triton is the only Moon that circles the planet in a retrograde rotation. The common theory for this occurrence is that Triton was a separate object captured by Neptune later in time.

PLUTO

For the longest time, Pluto was considered the ninth planet in our solar system. However, after the growth of technology and the discovery of similar worlds, it was reclassified as a dwarf planet.

ORBIT AND ROTATION

Pluto has the most eccentric orbit than all the other planets in the solar system: it is elliptical and tilted. One orbit around the Sun takes about 248 Earth years to complete. One day on Pluto lasts about 153 hours and has a 57° tilted axis. This means that Pluto rotates almost entirely on its side. Additionally, Pluto exhibits retrograde motion, similar to that of Uranus and Venus.

COMPOSITION

Pluto is slightly smaller than Earth's Moon and probably has a rocky core surrounded by a mantle of water and ice. Its surface consists of frozen methane and nitrogen, resulting in it having quite a low density. Additionally, Pluto has a large variety of formations on its surface, ranging from mountains to craters. These mountains, which are made of big blocks of water-ice, can be vast with a coating of frozen gases such as methane. Pluto's surface is dotted with craters, which show signs of erosion and filling. This determines the possibility that there might be tectonic forces might be at play.

ATMOSPHERE

Pluto has a thin atmosphere, which expands or contracts as it gets closer or farther from the Sun, similar to a comet. As Pluto gets closer to the Sun, the planet's surface evaporates, resulting in a "thicker" atmosphere. On the other hand, as Pluto gets farther from the Sun, the atmosphere might cool and fall to the surface in snow. Its composition is mainly that of nitrogen;

however, methane and carbon monoxide have also been detected. Additionally, due to Pluto's low gravitational pull, its atmosphere is largely extended to the planet's surface.

MOONS

Pluto has five moons: Charon, Nix, Hydra, Kerberos, and Styx. The common theory is that this moon system results from Pluto colliding with a similar-sized body sometime in its past. Charon is the largest of all the moons and is around half the size of Pluto itself. Due to its size, Charon is commonly referred to as the double planet. In comparison to Charon, the other moons are much smaller and irregularly shaped. Additionally, these moons are not tidally locked to Pluto and spin without keeping the same face to Pluto.

PLANET X

In 2015, Caltech scientists found evidence, through computer simulations and mathematical modeling, of the possibility of another planet in the solar system. There is heavy debate about its existence within the scientific community because this planet has not yet been directly observed.

Still, it's hypothesized that Planet X would have a mass about ten times that of Earth and has an orbit about 20 times further than Neptune. This orbit would take this planet about 10,000 or 20,000 years to make one full orbit around the Sun.

Its discovery was based on specific small, icy objects and dwarf planets clustering together in the Kuiper Belt. This unusual behavior was explained using the existence of such a planet.

ASTEROID BELT

Scattered in orbits around the Sun are particles of rock left from the dawn of the solar system. These objects orbit between Mars and Jupiter, thereby separating the terrestrial planets and the gas giants. Though this belt is huge, the belt's total mass is less than the Moon.

KUIPER BELT

The Kuiper belt is commonly confused with the Asteroid belt; the main difference is location. While the Asteroid Belt is located between Mars and Jupiter, the Kuiper belt exists beyond Neptune's orbit. The Kuiper belt is inherently a donut-shaped region of icy bodies beyond the orbit of Neptune. The belt contains millions of icy objects referred to as Kuiper Belt objects (KBOs) or trans-Neptunian objects (TNOs).

Like the asteroid belt, the Kuiper Belt is also created from the leftovers of the solar system's early creation. Additionally, the Kuiper Belt shouldn't be confused with the Oort Cloud, a much more distant region of icy bodies. The Kuiper Belt is the final frontier in our solar system and is still being extensively studied.

NEARBY STARS

There are around 133 bright stars, similar to the Sun, 50 light-years away from it. Most of these stars are visible with the naked eye. Within such a space, there may be around 1400 star systems in this space and approximately 2000 stars. Some intriguing stars are described below.

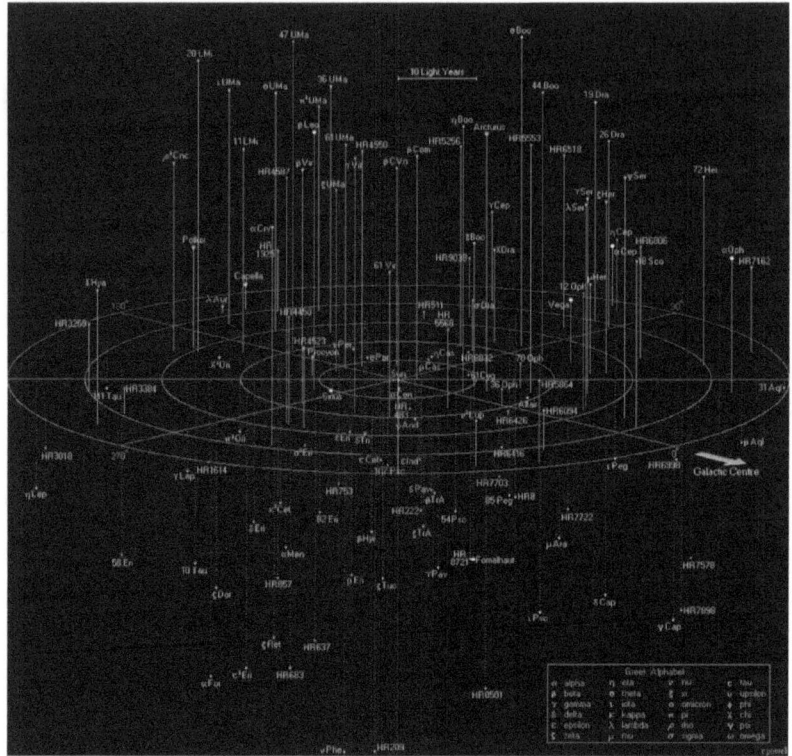

FIGURE 104: NEARBY STARS IN 50 LIGHT YEARS RADIUS [39]

ALPHA CENTAURI STAR SYSTEM

The closest stars to Earth are the three stars in the Alpha Centauri system, the two main stars being Alpha Centauri A and Alpha Centauri B, which form a binary pair. These stars are 4.3 light-years away from Earth. Both these stars orbit each other with a common center of gravity every 80 years. The average distance between them is around 23 AU, a bit more distance than the Sun and Uranus.

In August of 2016, astronomers detected an Earth-sized planet in the Alpha Centauri system. This world is known as Proxima B and is about 1.3 times more massive than the Earth. This planet is also in the star's habitable zone and in the right area where liquid water can exist. The planet is quite close to its host star and completes one orbit every 11.2 Earth days, and is likely tidally locked, meaning it always shows the same face to the host star.

So, one side of the planet is frigid, and the other to be incredibly hot. Though more studies need to be conducted on the habitable nature of Proxima B, it is interesting to know that life could exist outside of our solar system.

Both stars, Alpha Centauri A and Alpha Centauri B, would shine as one to the naked eye. Yet, through a telescope, it can be observed that there are, in fact, two separate stars. Alpha Centauri A is the third brightest star in the sky, while Alpha Centauri B is an orange K2-type star (slightly more massive than the Sun). Though difficult to see with the naked eye, another star in the Alpha Centauri system is Proxima Centauri.

PROXIMA CENTAURI

Proxima Centauri, unlike Alpha Centauri A, is too faint to see unaided in the open sky, and through a telescope, it seems quite far away from the other two stars. Proxima Centauri is a red dwarf about seven times smaller than the Sun and slightly larger than Jupiter. It is also the star that is rotated by the planet Proxima B, which may house life. Out of the three stars, Proxima Centauri is considered the closest to our Sun specifically, only about 4.22 light-years away.

BETA GEMINORUM

Beta Geminorum also known as Pollux, is a star that lies in the constellation Gemini. The star is a red giant that has finished the beginning of its life cycle and is fusing lighter elements into heavier ones at its core. The star is quite close to Earth, about 35 light-years away, and a luminosity of about 32 times the Sun.

It was recently discovered that the star has an exoplanet with a mass at least 2.3 times that of Jupiter and an orbital period of 590 days. The planet is known as Beta Genorium B was discovered by astronomers in 2006 using the Doppler method. Scientists observed that there were gravitational wobbles in Pollux that were explicitly being produced by the planet. There is little known about the star and planet; further research is being conducted on the possibility of life in that part of the universe.

ZETA RETICULI

Zeta Reticuli has been widely claimed as the "alien star system," specifically becoming famous for the Betty and Barney Hill story. Now, I am entirely unsure whether these claims are valid, and through the possibility of aliens is a real probability, that is not something we will be debating. Instead, we will be discussing an overview of the Zeta Reticuli star system since it is one of the closest ones to our own.

Zeta Reticuli is a binary star system located in the southern constellation Reticulum. The system is composed of two Sun-like stars: Zeta and Zeta Reticuli. Both these stars are, just barely, visible to the naked eye in areas without light pollution. It is known that these stars rotate around a common center of mass and compose the majority of the system, as no planets have been found. For a world to be in a habitable zone of Zeta Reticuli, it would need to be at a distance of 1 AMU. Therefore the planet is similar to the Earth and provides it an orbital distance that would take 320 days to complete.

The Zeta Reticuli system is part of the Zeta Herculis Moving Group. All the group members formed in the same molecular cloud and have known to move as a unit around the Milky Way. Like all other star systems, Zeta Reticuli orbits through our galaxy.

Finally, though Zeta Reticuli has been a famous name in literature and UFO circles, it is still unknown whether such a thing is real; only the future can tell.

GALAXIES

When gazing into the night sky, you might notice hundreds and thousands of stars all over the sky. Interestingly, though you don't realize it, the majority of these stars are imposters. Most of those points of light are, in fact, galaxies.

Galaxies are collections of millions or trillions of stars composed of stars, dust, and dark matter, all held together by gravity. Astronomers are still unsure of how galaxies are formed. The common theory is that gravity pulled dust and gas together to form individual stars after the big bang. Those stars drew closer together into collections that became galaxies.

Additionally, in 1924 Edwin Hubble was the first to notice that galaxies are moving farther apart from one another. He measured the distance between individual galaxies using the Doppler shift (how much light was stretched from galaxies due to their motion). Overtime, Hubble observed that the galaxies around the Milky Way were moving away at incredible speeds and thereby determined that the universe is expanding.

Most galaxies are absorbed to have black holes at their centers and can produce tremendous amounts of energy. At times, the central black hole of a galaxy can be incredibly active for even smaller galaxies. Galaxies are classified by their shape. Some galaxies, such as the Milky Way, have arms spiraling outward from their center. These are known as spiral galaxies and are considered the majority of galaxies that astronomers have observed. The gas and dust in spiral galaxies circle the center at speeds of hundreds of miles per second, creating a pinwheel shape. There are also galaxies known as "barred spirals," with a barred structure in the center formed when dust and gas are funneled in the center.

Elliptical galaxies are similar to spiral galaxies; but, they lack spiral arms. Their appearance ranges from circular to ellipses. These galaxies have less dust than their spiral cousins, so the star-making process is minimal, making most of the stars in this galaxy older. Though they make up a small portion of visible galaxies, astronomers believe that over half of the galaxies in our universe are elliptical.

The remainder of the galaxies are known as irregular galaxies. They have neither round nor spiral, and there is a lack of a general shape at all. These galaxies lack definition and are usually affected by the gravity of other galaxies; this results in them becoming stretched or warped. Collisions can also affect their shape leading to deformation.

Galaxies don't stand still in space. They are bunched together in groups known as clusters. Some clusters are quite large and contain thousands of galaxies, while others are much smaller. Occasionally, galaxies in clusters slam into one another, thereby merging into one. During galactic collisions, individual stars don't collide with another. However, the influx of gas and dust leads to a large bump in star formation. The Milky Way is set to collide with the Andromeda galaxy in about a couple of billion years. Collisions are an important step in the life cycle of a galaxy, making it an imperative step.

MILKYWAY

Every star that one sees in the sky is part of a galaxy known as the Milky Way, a vast spiral galaxy stuffed with between 100-400 billion stars, with planets of their own. Even our solar system resides inside of this galaxy. The Milky Way got its name because of how it looked from Earth, like a streak of milk across the sky. This streak was a hazy white band made up of stars, dust, and gas. Since we are inside the galaxy, it looks like a stripe. Now, if we were to look at it from above the galaxy, it would look like an enormous spiral about 200,000 light-years across.

The spiral is made out of four large arms of stars connected by a bar in the center of the galaxy. The arm that houses the Earth is called the Orion arm, as is one of the smaller arms, where our solar system resides 26,000 light-years from the center of the galaxy. Our galaxy is also rotating, where it takes us about 240 million years to complete one circle around its middle.

It is unknown what is at the center of the galaxy. Yet, scientists theorize that it's a supermassive black hole known as Sagittarius A. This black hole has a mass of more than 4 million times the Sun and is cramped into a tiny diameter. It is theorized that there are black holes at the center of most galaxies. We will discuss this further in the black hole section.

NEARBY GALAXIES

Astronomers understand that the Milky Way isn't alone in the universe. There are hundreds of thousands of galaxies all over the universe. However, a collection of 54 galaxies and dwarf galaxies make up our Local Group. Additionally, our galaxy is also part of a more extensive formation known as the Virgo Supercluster.

A common misconception is that the closest galaxy to us is the Andromeda Galaxy. It is the nearest spiral galaxy. Not necessarily the closest galaxy, not by a long shot. There are a couple of dwarf galaxies around us that are much closer.

CANIS MAJOR

The closest known galaxy to the Milky Way would be Canis Major, which is a dwarf galaxy. From our solar system, this galaxy is a mere 25,000 light-years away. Overall, this galaxy is believed to contain one billion stars in total, most of which are in the Red Giant Branch. In addition to being a dwarf galaxy itself, there is a long filament of stars trailing behind it. This complex, ring-like structure is referred to as the Monoceros Ring and wraps around the galaxy three times.

Multiple other globular clusters orbit the center of our Milky Way as satellites. These are thought to have been originally part of the Canis Major Dwarf Galaxy, yet, formed due to the lack of a gravitational pull from the dwarf galaxy. Before the discovery of this galaxy, astronomers believed the Sagittarius Dwarf Galaxy was the closest to us. But, it is still a neighbor.

SAGITTARIUS DWARF GALAXY

Between 300-900 million years ago, our Milky Way galaxy nearly collided with the Sagittarius dwarf galaxy. According to data collected from the ESA's Gaia mission, there are constant ongoing effects of these events. For example, stars are moving like ripples on the surface of a pond. Additionally, this galactic collision is also part of the ongoing cannibalization of the dwarf galaxy, the Milky Way.

Located about 87,000 light-years away lies the dwarf galaxy Sagittarius. The reason for its name is because it's located in the direction of the Sagittarius constellation. As stated previously, there are several dwarf globular star clusters following the Milky Way galaxy. Originally, these clusters were considered part of the Milky Way that broke off; however, it was discovered that these are part of the Sagittarius Dwarf Galaxy.

Initially, the Sagittarius Dwarf Galaxy was considered to be a globular cluster of stars. It was a very faint nebula, discovered near the Sagittarius constellation. Over time, as the study of the globular clusters that trail the Milky Way, specifically, M54, it was noted that this cluster seemed to rotate around a separate nebula. Upon further investigation, it was discovered this was a separate dwarf galaxy and was considered to be the closest to the Milky Way before the discovery of Canis Major.

ANDROMEDA GALAXY

The Andromeda galaxy is consistently viewed as the closest neighbor to the Milky Way galaxy, though it is, in fact, the nearest spiral galaxy to the Milky Way. Additionally, it has also been noted that the Andromeda galaxy is steadily headed towards our course with the Milky Way. Luckily, this won't be happening for the next few billion years.

From our solar system, Andromeda is about 2.5 million light-years away, and like the Milky Way, it also contains a bulge of matter in the middle. It has a similar structure, with the bulge being surrounded by a disk of gas, dust, and stars in a halo. The Andromeda galaxy has been observed to contain approximately a trillion stars, more than the 250 billion in the Milky Way. Our galaxy is more massive because it is thought to contain large amounts of dark matter.

A collision between the Andromeda and the Milky Way galaxy would alter the structure of both galaxies immensely. Scientists predict that the collision would happen about 4 billion years from now. By that time, the Sun would have run out of fuel and swallowed the Earth (that gives us other things to worry about). During collisions, there is a messy phase where the arms project erratically from the combined pair; yet, the two will settle into a smooth elliptical galaxy. Additionally, galaxy collisions are a normal part of the evolution of the universe. Both the Milky Way and Andromeda have shown signs of collision with other galaxies in their past.

BLACK HOLES

There is very little officially known about black holes and their properties. In general, black holes are known to be volumes of space where gravity is extreme enough to prevent even light from escaping. German physicist Karl Schwarzschild proposed the most modern concept of a black hole in 1915 after determining the exact solution of Einstein's approximations of general relativity [40].

This equation helped Schwarzschild realize that there was a possibility for mass to be squeezed into infinitely small points. This extreme gravitational effect would cause spacetime to bend around this particle where not even photons could escape its curvature. The cusp of a black hole, right before the complete slide into oblivion, is known as the event horizon. The distance between the event horizon and its infinitesimal core is named after Karl Schwarzschild. The event horizon is where the escape velocity equals the velocity of light. Outside of the horizon, the escape velocity is less than the speed of light, so there is the hope of escaping a black hole's event horizon. Also, there is no way to escape if one finds themselves inside the horizon.

Suppose you throw a rock up into the air. Assuming that you didn't throw the stone too hard, it will fall back down due to Earth's gravitational pull. Now, if you were superhuman and threw the rock hard enough, you could have it escape the Earth's gravitational pull entirely. The velocity required to escape a planet's gravity is known as the escape velocity. This velocity increases as the mass of the planet increases. More massive planets have a higher escape velocity due to the increased gravitational pull associated with their mass. Within a black hole, the mass is so concentrated that the escape velocity is greater than the velocity of light; therefore, nothing can escape the object's gravitational field.

FORMATION OF A BLACK HOLE

It has been known that when stars run out of fuel, they undergo an extreme gravitational collapse. This incredible amount of mass in such a small confined area results in an extreme density, resulting in a black hole. Not all stars become black holes. The star's mass has to be considered at least three times greater than that of our Sun. Only then will the star undergo a complete gravitational collapse. This type of black hole is one that is most commonly observed; yet, there is another that is hypothesized.

This hypothesized black hole rapidly formed when a rippling vacuum of the universe's beginnings rapidly expanded in an event known as inflation. During this time, highly dense regions collapsed and became primordial black holes that would be slightly bigger than a jelly bean.

INSIDE THE BLACK HOLE

Since nobody has ever been inside a black hole, there is no way to know what happens past the event horizon. Yet, several theories describe the possibilities inside of there. First, the horizon has some interesting geometric properties. The horizon seems to be an ice static spherical surface to an observer, watching the black hole from some distance away. But once you get close, one realizes that it has an enormous velocity constantly swirling around the black hole. This geometric illusion explains why it's easy to cross the horizon in that inward direction. But, almost impossible to get out. The horizon seems to be moving out at the speed of light, and to escape; one would have to travel faster than the speed of light (all of this might sound incredibly strange, don't worry. It still perplexes scientists to this day)

If you manage to get inside the horizon, spacetime distorts so much that a spaceship starts moving forward through time or into the future. Avoiding the center of a black hole becomes similar to avoiding, next week, there is no physical way to do so. Eventually, you would hit the singularity. Singularity is at the center of the black hole where density and gravity become infinite and spacetime curves forever, and the laws of physics cease to operate. There is honestly very little known about the intricacies of singularity, and most of it remains theoretical since nobody has been inside a black hole.

OUR EARTH

One of the significant aspects of our world today is, well, our Earth. There are several different parts of Earth, ranging from the atmosphere to the inner core. All of these different layers have a variety of physical properties, which are discussed below [41] [42].

ATMOSPHERE

The atmospheric properties of a planet help determine a variety of different aspects of it. For example, the discovery of small amounts of phosphine gas in the atmosphere of Venus determined the probability of life existing in its upper atmosphere. So, the Earth's atmosphere also indicates a large aspects of it.

TROPOSPHERE

The troposphere is the lowest level of Earth's atmosphere, where the temperature decreases as the altitude increases. The tropopause is the top of the troposphere, which occurs at an altitude of 8 km at the poles and 18 km at the equator. The decreasing gravitational pull of the Earth results in the atmosphere's density to fall exponentially as the height increases. The troposphere contains 80% of all the mass and most water vapor in the entire atmosphere, resulting in clouds and stormy weather.

As height increases, the temperature decreases at a constant rate due to a phenomenon known as vertical mixing. The air near the surface is heated, leading to it rising. As it rises, the air temperature cools and becomes heavy, therefore sinking. This is the process of convection. Convection relaxes the temperature profile toward a stable configuration. This stable configuration is known as the adiabatic temperature lapse rate, for which there is a constant decrease in temperature for the decreasing pressure (e.g., the increasing height).

OZONE LAYER

The ozone layer is a section in the stratosphere where 90% of all the Earth's ozone is found. Ozone absorbs the most energetic wavelengths of UV light, known as UV-C and UV-B. The oxygen molecules, also located in the layer, absorb other forms of UV light. Together, the ozone and oxygen molecules can absorb 95-99% of the UV radiation that reaches our planet. Ozone and oxygen molecules are continuously being formed, destroyed, and re-formed due to the constant bombardment of UV radiation from the Sun. This UV radiation breaks the bonds between atoms, creating free oxygen molecules that are highly reactive. If a free oxygen atom bumps into another oxygen molecule (O2), it forms ozone (O3) due to its high reactivity.

In 1976, a British scientist at Halley Bay started recording low amounts of ozone above Antarctica. These scientists detected a 10% drop in ozone levels since 1957, specifically during the Arctic spring (September - November). This was the first discovery of the ozone hole, which is not necessarily a hole, but the thinning of the ozone layer that changes seasonally. In 1985, scientists determined this hole to be human-made. Since the 1960s, chlorofluorocarbons (CFCs) started being used to manufacture air conditioners, aerosol spray cans, Styrofoam, and cleaning products. CFCs can break apart ozone molecules due to the chlorine molecule released when a CFC is bombarded with UV radiation.

This ozone hole increases or decreases depending on the season. During the dark winter, the airdrops to temperatures resulting in icy clouds forming. Reactions on these icy clouds' surface release chlorine from CFCs, which reacts with ozone, destroying it. Countries are trying to work together to address ozone destruction, and hopefully, we will see improvements soon.

MESOSPHERE

The mesosphere is located between the stratosphere and the thermosphere, making it incredibly challenging to study. Weather balloons and other aircrafts are unable to reach it, and satellites orbit above the mesosphere, thereby making it incredibly difficult to measure that layer's traits. Commonly, scientists use instruments on sounding rockets to measure the mesosphere directly; however, these flights are infrequent and too brief to take detailed measurements. Overall, this results in the mesosphere being a bit of a mystery even today.

What is known is that most meteors vaporize in the mesosphere. Therefore, that layer has a high concentration of iron and other metal molecules or atoms. Like the troposphere, near the top of the mesosphere, the atoms of different molecules separate into groups. But it is found that near the bottom, these same molecules converge and collide due to the layer's volatility.

Interestingly, strange, high altitude clouds known as "noctilucent clouds" form near the poles. The reason for these clouds being peculiar is that they are higher up than other types of clouds. Additionally, the mesosphere is much drier than the troposphere, making it even surprising that clouds can form in that area. Finally, the mesosphere can influence waves and tides, therefore driving most of the Earth's global circulation.

IONOSPHERE

There are many regions in Earth's atmosphere that have a large amount of electrically charged atoms and molecules. The atmosphere is commonly bombarded with high energy UV or X-ray waves. These collide with molecules and atoms, leading to electrons freeing themselves from atoms and becoming charged ions. These ions and electrons move very differently in comparison to neutral atoms and molecules. Areas in Earth's atmosphere where there is a high density of such ions is known as the ionosphere. Unlike the other layers of the atmosphere, the ionosphere isn't a distinct layer. Instead, the ionosphere is specific areas within the original layers of the atmosphere that are highly electrically charged.

In comparison to other layers, the ionosphere is very malleable. It varies mostly depending on the time of day. During the day, UV and X-ray radiation continuously knock electrons out of atoms and molecules, producing a constant supply of free electrons and ions. At night, these electrons merge with atoms and molecules to form neutral particles. This process is much more consistent at night when there isn't a constant bombardment of UV/X-ray radiation from the Sun, thereby decreasing the overall number of ions in the atmosphere.

Like how the ionosphere changes depending on the time of day, it is also affected by seasonal changes. Depending on the time of year, the chemistry of the cosmosphere also changes, affecting the ionosphere as well. The solar cycle is also an essential aspect of the number of ions in the atmosphere. Depending on the brightness of the Sun, the composition of the ionosphere changes rapidly. So, large geomagnetic storms triggered by the solar flares from the Sun can cause temporary disruptions to the ionosphere.

One of the significant uses of the ionosphere is radio communication. Some frequencies of radio waves (short waves) bounce off electrically charged particles in specific ionosphere areas. Radio communication takes advantage of this phenomenon and uses the ionosphere to increase the range of these signals. Before satellite communication, radio operators had to deal with disruptions within communication due to changes in the ionosphere. The ionosphere can absorb, dampen, bend, or reflect radio waves, leading to these waves' accuracy changing over time.

SCHUMANN RESONANCE

The Schumann Resonance, a global electromagnetic resonance phenomenon named after physicist Winfried Otto Schumann, has been the researchers' eager study in recent years. This phenomenon shows that Earth acts like a giant electrical circuit, where Earth's atmosphere acts like a conductor. The ionosphere is a region where, due to solar radiation, the individual electron is dislodged from neutral gas atoms. This process creates positively charged ions, making the ionosphere highly conductive and able to trap electromagnetic waves. Also, Ionosphere in our atmosphere functions as an electrically charged cavity.

Between Earth's surface and the ionosphere, there is a section containing a total electrical charge of 500K Coulombs. There is a current flow between the ground and the ionosphere, which has a voltage potential of 200K volts. Additionally, the atmosphere has a resistance of approximately 200 Ohms. All this means that there is a large amount of electrical activity between the Earth's surface and the ionosphere. This activity is also in the form of standing waves of electricity known as the Schumann Resonances. In other words, the atmosphere is continuously resonating at a radio frequency of 7.83 Hz, which is the numerical representation of the Schumann Resonance. The amount of resonance fluctuates as the ionosphere becomes dense. The density depends on the amount of solar radiation striking it.

THERMOSPHERE

Although the thermosphere is considered part of the Earth's atmosphere, the air density is so low that it mimics the temperature and pressure of outer space. The air density is so low that most of the satellites and space shuttles orbit within the thermosphere.

Due to the low air density, unlike the previous layers, gases in the thermosphere collide so inconsistently that they become separated. This results in much of the Sun's electromagnetic radiation breaking apart in this layer. The Sun is very active; it releases large amounts of EM radiation towards the Earth; this results in the thermosphere expanding or "puffing up". Due to this phenomenon, the height of the top of the thermosphere varies with time.

Additionally, high-energy solar photons tear electrons away from the separated gas molecules in the thermosphere. Creating many electrically charged ions that collide with other atoms and molecules, exciting them into higher energy states. However, when these ions release energy and fall back down to a lower energy state, the output is light, thereby creating the beautiful Aurora (Northern and Southern lights).

EXOSPHERE

Not all scientists agree whether the exosphere is a part of the atmosphere. Since the exosphere is extremely thin and gradually fades into outer space, there is no clear upper boundary of this layer. Since the air is incredibly thin, there are very few collisions between atoms and molecules in this region of the atmosphere. Gas atoms and molecules move along "ballistic trajectories" as they curve back towards the Earth under the influence of gravity. Specific fast-moving particles don't return and fly into space. This is how some of the atmosphere leaks out into space.

Many satellites orbit below and within the exosphere. Even though the air is incredibly thin in these layers, there is still a bit of drag force that ensures satellites can orbit. Over time if there is no boost given, the satellite will re-enter the Earth and burn up, hence why there is periodic an upward boost given to these space satellites.

MAGNETIC FIELD

Unlike most other planets in our Solar System, the Earth is like a giant bar magnet. The north pole of this magnet is located near the top of the planet, and the south pole of the magnet is situated near the geographic south pole. Interestingly, if you could draw a line between the magnetic south and north poles, you would find that the magnetic axis is tilted 11.3° away from Earth's axis of rotation.

The current theory for Earth's magnetic field is generated by electrical currents flowing in the liquid outer core. The liquid metal deep within the Earth moves through convection currents. It's believed that these movements set up currents and the magnetic field.

One of the many important aspects of Earth's magnetic field is that it helps block space radiation from hitting us. The most common form is the solar wind, which are highly charged particles and ions blasted out of the Sun, similar to a steady wind. Earth's magnetic field channels and moves the solar wind around the planet so that it doesn't affect us. Over time, without the magnetic field, space radiation would strip away our atmosphere, similar to what occurred to Mars.

Earth's magnetic poles are known to move up to 15 km around the surface. This is due to the consistent magnetic field reversal that occurs approximately every 250,000 years. The north magnetic field becomes the south and vice versa. As of now, there is no apparent reason why this reversal happens. Interestingly, we are long overdue for a reversal. The last one happened about 780,000 years ago.

VAN ALLEN BELTS

In 1958, James Van Allen discovered giant donut-shaped swaths of magnetically trapped, highly energetic charged particles. Upon launching the Van Allen probes in 2012, our understanding and observations of these belts have enhanced. These areas are incredibly electrically charged, leading to them causing issues among satellites and space ships that cross this area. Hence why leaving the Earth becomes so tricky, considering how these belts negatively affect communication and the systems within the ship.

INNER LAYERS

The layers of the Earth under our feet or inner layers of the Earth are as follows.

CRUST

The crust is the thinnest layer and the topmost area of the Earth. It is the landscape on which we live. Due to the convection currents in the mantel, the crust is broken up into areas called plates. These tectonic plates move around and converge, leading to mountains or rifts all over the Earth's surface.

Continents are composed of relatively lighter blocks that float on the mantle, like icebergs. On the other hand, the seafloor is made of denser rock that presses more directly against the mantle.

MANTLE

A common misconception about the mantle is that it consists of lava; but, it is rock. This rock is so hot that it flows and moves under pressure, creating slow-moving convection currents, like those of the outer core.

OUTER CORE

Above the inner core is the outer core, which is a shell of liquid iron. Due to the iron being in a liquid state, it is continuously moving due to convection currents similar to boiling water. This movement results in the creation of Earth's magnetic field. Like the inner core, it is mainly composed of iron with various other elements in smaller amounts.

There are some inconsistencies in our understanding of the inner and outer cores. As we know, Earth has a magnetic field, and it is explained by the presence of iron in the inner and outer cores. This theory may not be right as iron loses its magnetic properties at high temperatures.

INNER CORE

The inner core is the last layer beneath the Earth's surface and is assumed to be made of a solid iron core. Although the inner core is incredibly hot, the pressure is so high that the iron is unable to melt. Yet, this iron is not pure. It also has a mixture of sulfur and nickel along with smaller amounts of other elements.

INTERESTING THEORIES OF PHYSICS

The main principle behind physics and science, in general, is to help explain all the phenomena we see in the Universe. Unfortunately, most of the information we have learned thus far does not fully explain the Universe we are witnessing today. So, these anomalies give rise to other newer potential theories. As we will describe in better detail, unification is a crucial aspect of physics, finding a unifying theory that would explain everything. One main problem within unification is explaining the connection between quantum mechanics and the general theory of relativity.

Quantum mechanics explains the behavior of atomic particles, while the general theory of relativity explains the behavior of massive bodies such as planets, stars, galaxies, etc. The theories described in this section will attempt to connect the micro and macro bodies. Many of these have not been proven; but, they provide a unique insight into what might truly be happening within our Universe.

STRING THEORY

Unification has always been a key factor in physics since its humble beginnings. Isaac Newton united the heavens and the Earth, showing that the same laws governed both the motion of planets and the trajectory of a spinning wheel. Around 200 years later, James Clerk Maxwell took unification to the next level, revealing that electricity and magnetism are two aspects of one force and can be described by a single mathematical formula. Einstein linked space-time, envisioning the grand synthesis of all of physics.

Ever since the late 20th century, physics has become more and more complicated. With ideas ranging from quantum mechanics to Einstein's theory of relativity, everything seems to become more disjointed. This perception, this dysconnectivity, and the need for unity led to the development of the string theory [43].

String theory attempts to find an overarching framework that can explain all physical reality by uniting two pillars of modern physics: quantum mechanics and Einstein's general theory of relativity. The main principle of string theory is that particles are one-dimensional strings whose vibrations determine the particle's properties. Additionally, these "strings" vibrate in 11 different dimensions that humans cannot detect due to their innately small size. A string of a particular length hitting a particular note achieves the properties of an electron. In contrast, another string folded and vibrating at a different frequency plays the role of a photon. Due to these complicated movements across so many dimensions, humans witness all matter in the universe as particles instead of strings.

To understand why these extra dimensions are hidden from view, imagine a tightrope walker on a wire. To the tightrope walker, the wire is a one-dimensional line, its length. Now imagine a colony of ants also on the same wire. To these ants, the rope exists in two dimensions, its length, and thickness. It's by this same logic that to humans, the world would be in three dimensions; to strings, the world would exist in 11 different dimensions.

The reason string theory remains so appealing among physicists is due to its mathematical parallels across a variety of distinct subjects. For example, the same mathematics used in string theory to describe quantum entanglement can also describe black holes. Therefore, giving it the nickname of "The Theory of Everything."

String theory assumes the existence of supersymmetry; this means at the basic level, two types of particles should exist: bosons and fermions. And there should be a fermion for every boson. In the standard model of atomic particles, there are 18 known particles. String theory expects another 18 more particles to complete the supersymmetry. String theory is highly mathematical, and so far, we have not found any physical proof. The theory assumes at small spaces, there are many dimensions, and when we study macro bodies, we fail to notice these dimensions due to their innate size.

Interestingly, string theory opens the possibility of multiple parallel universes. The universe is known to have originated from a giant explosion that is known as the big bang. This theory brought up the idea of inflation in the 1980s. Inflation is a period of super-fast, accelerated expansion in early cosmic history. At the end of inflation, the energy ignites and explodes into the big bang. Scientists speculated that the end of inflation is triggered by quantum particles, thereby opening up the possibility of new universes constantly forming, separate from our own. Coupled with string theory, not only can new universes be forming in our dimension, but in a multitude of different dimensions, thereby opening up a multiverse landscape.

The string theory framework has been used to explain various phenomena ranging from the entropy of a black hole to gravity. However, this framework has a variety of issues when addressing the expanding universe. In the 11 other dimensions described by string theorists, there is no equivalence between force and matter particles, necessary in the real world. Additionally, there is no fundamental way to prove the existence of such "strings" in place of particles, or at least modern technology has not progressed to that extent.

PROBLEMS WITH ASSUMPTIONS

Commonly, scientists try to explain the behavior of the universe and atomic particles with the existing assumptions depicted by the laws of physics. Another school of thought mainly supported by Prof. Konstantin Meyl from the University of Applied Sciences in Furtwagen, Germany, gives rise to newer theories that break and remake previous assumptions. He advocates for more contemporary theories and some modifications to existing theories based on the concepts of scalar waves, vertices, and fields. Some of the ideas described in these sections are based on his explanations for the universe.

EXPANDING UNIVERSE

"The Big Bang" is considered the start of everything. A giant fiery explosion gave birth to planets, stars, and us. This explosion has now continued into today, and the Universe is a forever expanding mass. This assumption of "the expanding Universe" goes upon the basic idea that the space between astronomical bodies is increasing. Edwin Hubble "discovered" this when analyzing data gathered on the Doppler effect in the Universe. The Doppler effect involves a shift in the color or property of light. When an object moves towards us, we see a blueshift and a redshift if the object moves away from us. We view a change in frequency of light when an object moves towards or away from us, which leads to this change of color.

As described by the special theory of relativity, if a heavenly body moves at half the speed of light and we from Earth measure the light being emitted from this object, it will always be constant no matter if the mass is moving towards or away from us. In other words, we don't see a change in the speed of light, no matter how the object is moving. The only thing we observe from Earth is a change in frequency of the light from a Doppler shift or the Doppler Effect.

As we know from the special theory of relativity, if a heavenly body is moving at half (or 50%) the speed of light and we measure the speed of light coming from the celestial body, it is always constant even if the object is moving towards us or away from us. Scientists don't measure the speed of light as 50% of the original value if the object moves towards us and 150% if the object moves away from us. The only thing we see is the change in frequency of the light in the form of the Doppler effect.

Another observation that we witness is within galaxies. It is commonly known that all stars and planets are slowly moving towards the black hole in the middle of the galaxy. Therefore, galaxies are slowly contracting. If a galaxy is contracting, it can be assumed that the distance between the stuff (planets, stars, nebulas, etc.) within the galaxy is decreasing. Hence, a blueshift should be witnessed through our observations of the stars in our galaxy. The actual observations do not depict any blueshift. Scientists try to explain this phenomenon by stating that these stars are fixed, and the distances are unchanging. However, this explanation contradicts the known fact that stars and planets are moving towards the center of the galaxy.

Additionally, only the stars in other galaxies show a redshift, even as those galaxies contract. Therefore, a redshift does not necessarily mean that the Universe is expanding. It can be due to the stars in those galaxies moving towards the center of the galaxy. The reality of only seeing a redshift and not seeing a blueshift is a mystery yet to be solved.

GALAXIES AND KEPLER'S LAWS

As described in a previous chapter (Kepler's Laws of planetary motion), planets move in elliptical orbits with a maximum speed near the Sun and lower speeds father away from the Sun. The basis for these laws is the force of gravity, which is the basis of all motion in the Solar System. Using the idea of gravity to explain the motion of stars in galaxies creates several issues, which have yet to be explained.

The shape of galaxies remains constant, like our Milky Way, which is a spiral galaxy. Additionally, the stars within our galaxy are not moving in elliptical paths around the central black hole, as Kepler's Laws would predict. Interestingly, stars closer to the center of our galaxy also move slower than those farther away, which is contradictory. The gravitational pull of the black hole grows the closer one gets to it. These issues give rise to one central question: how are galaxies maintaining their shape? Physicists have postulated the existence of dark matter to explain this strange behavior. Yet, there has been no proof found for its actual existence. Hence, the true physical properties of galaxies remain a mystery.

MAXWELL'S EQUATIONS

As we described previously, Maxwell's second equation states that the divergence of a magnetic field is zero (magnetic monopoles don't exist). In simpler terms, we will always find magnets with north and south poles together. But, as hinted on previously, recently, researchers from Helmholtz-

Zentrum Berlin, Germany, observed the existence of magnetic monopoles. Therefore, the value of Maxwell's equation is nonzero, thereby making us question and revisit many of the original theories that assume the nonexistence of magnetic monopoles.

So, Maxwell's equations we use today have this error. Additionally, another mistake within the derivation of these formulas was the scalar constant. When these equations were first proposed, they were unable to prove the existence of scalar waves, so this constant was mistaken as zero within the original formulas. In effect, Maxwell's equations for electromagnetism consists of both transverse and longitudinal waves. In general, these mistakes open up some consequences and makes us re-evaluate the assumptions that we've made thus far.

FIELDS AND DISTANCES

The following concepts are based on Prof. Konstantin Meyl and provide an interesting insight into modern theories and laws. As we learned earlier, both electric and magnetic field intensities are directly proportional to the inverse square of the distance. So, if you were to half the distance, the intensity would quadruple. This simple inverse square law is the basis of the entire subject of electromagnetism.

If we were to make a general assumption that any field upholds this law, then the field itself would begin defining the concept of distance. To better understand this concept, let's assume that the entire universe is surrounded by a field known as aether. By the logic stated previously, all distances and speeds inside the field would be defined by the field.

We learned earlier that both the electric field and magnetic field intensities are proportional to the inverse square of the distance. Incidentally, the entire subject Electromagnetism and also waves, are based on this inverse square law. Suppose now we make a general assumption that this law holds for any field, i.e., the field intensity is proportional to the inverse square of the distance, then the field itself defines the concept of distance. Assume aether is also a field, then distance and speed are defined by the aether field. As we measure the speed of light as constant in all known situations in the vacuum, this may be because the distance (length of measurement) itself is changing in the aether field such that the value of the speed of light remains constant. So, the speed of light is observed as constant, but the actual distance and time might be varying. This phenomenon with light might be happening because the measuring instrument is affected by the field changes of aether, which produces a constant speed of light. Proving or disproving the properties of aether will be extremely difficult now, and maybe in the future, we will be able to discover some new information.

VACUUM ENERGY

Think of the entire universe as a swimming pool where the fabric of space is the water inside the pool. Like how one would take buckets of water out of the pool, scientists speculate that there is a possibility to take "buckets" of energy out of the emptiness of space itself. This "water" or background energy is known as vacuum energy and exists throughout the entire universe. The existence of this type of energy has been formulated in the Heisenberg Uncertainty Principle, and its effects have been observed (spontaneous emissions, Casimir effect, Lamb shift, etc.). Some scientists believe that the fabric of space is an extremely dense fluctuating of electric and magnetic fields, thereby allowing the possibility to harness limitless quantities of energy from the emptiness of space. Yet, the existence of vacuum energy is not fully understood or explained.

ZERO POINT ENERGY

Zero-point energy is a particular case of vacuum energy. It is thought the fabric of space is actually "quantum foam" with extremely dense (10^{94} g/cm3) electric and magnetic fields. After constantly being studied since the discovery of quantum mechanics in the 1920s, there can be no doubt that ZPE (Zero Point Energy) or vacuum energy is a real physical effect. It is, in fact, an interesting unavoidable part of quantum physics. Observations indicate that it is also possible to manipulate this energy since any objects that change vacuum energy (electrical conductors, gravitational fields, etc.) also distort the quantum mechanical vacuum state.

SPACE RADIATION

Easy space travel is always considered the dream for most physicists. However, many hurdles need to be crossed before traveling to Mars becomes like a vacation choice. One of these many hurdles is space radiation. There are two primary types of space radiation: non-ionizing or ionizing. Ionizing radiation consists of particles that have enough energy to remove an electron from its orbit.

On the other hand, non-ionizing radiation doesn't contain enough energy to remove electrons from the material it crosses. Though non-ionizing energy is also damaging, it can be easily shielded out of an environment. Ionizing radiation is difficult to avoid because it can move through substances and alter them as it passes through. Ionizing radiation can be viewed like an atomic cannonball that blasts through the material leaving large amounts of damage behind.

COSMIC RAYS

Space radiation consists of three different types of radiation: particles trapped in Earth's magnetic field, particles shot into space during solar flares, and galactic cosmic rays (which are high-energy protons and trace ions from outside our solar system). Intrinsically, space radiation is composed of atoms in which electrons have been stripped away, and they accelerate in space at speeds close to that of light. Over time, only the nucleus of the atom remains. This form of radiation is known as ionizing radiation.

Galactic cosmic radiation is the most dominant source of radiation in space. GCR is formed of the nuclei of atoms that have their surrounding electrons stripped away, and their traveling at approximately the speed of light. These particles can cause the atoms they pass through to ionize and consistently pass unimpeded through a typical spacecraft or an astronaut's skin.

Therefore, electrons become a serious issue when considering space travel. It has consistently been known that the Van Allen belts around the Earth are filled with this form of radiation. There has been a lot of fear when traveling to Mars due to this radiation harming an astronaut's brain. New mechanisms need to be put in place to prevent this from further hindering space travel.

NEUTRINO RADIATION

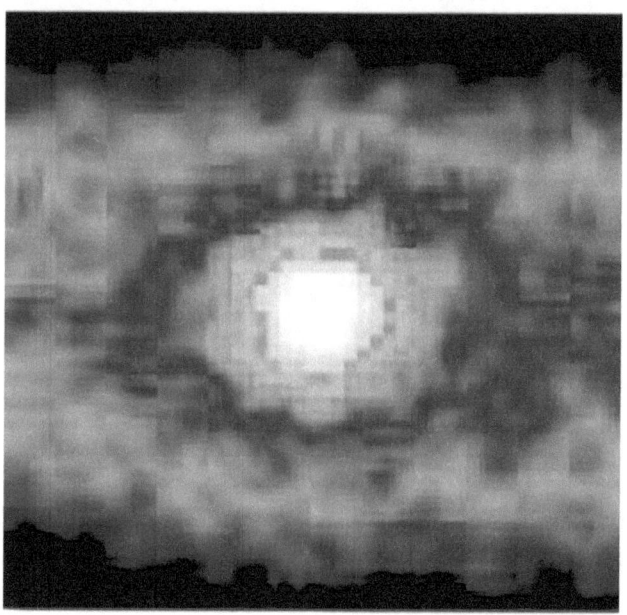

FIGURE 105: SUN IN THE MORNINGS WITH NEUTRINO RADIATION [44]

Neutrinos are small atomic particles with a negligible mass, ½ spin, and are positive, negative, or no charge. They travel faster than light and pass through ordinary matter and atoms, interacting very minimally with water and other substances. Neutrinos are produced in beta decay from stars, black holes, supernovas, cosmic rays striking atoms, nuclear reactors, nuclear bombs, particle accelerators, and other heavenly bodies.

The majority of the neutrinos found on Earth are from the Sun. The neutrino radiation from the Sun is massive (6.5×10^{10} neutrinos per second per square centimeter). This invisible neutrino radiation is very intense and, like electrometric or light waves, propagating from the Sun. The neutrino radiation which propagates from the Sun is half the amount at night than in the day. This surprising difference suggests that Earth is absorbing or interacting with half the neutrinos coming from the Sun. Based on this neutrino behavior, it has been postulated that Earth is expanding and increasing in size slowly (approximately 29 cm circumference per year). It is also believed that neutrinos interact with water; hence all creatures on Earth might be affected by the neutrino radiation; but, more research still needs to be done.

HARVESTING NEUTRINO ENERGY

Fossil fuels are not going to be the future energy production system. So, scientists are always trying to determine new ways to solve the current energy crisis, and neutrinos might be one solution. Harvesting neutrino radiation on Earth might solve all our energy problems. This radiation is freely available in abundance, and at present, we are not using it. This radiation's use was found by chance while building the Rail Gun. The Rail gun is claimed to work based on electromagnetism and Lorentz force, but it also uses neutrino radiation from the Sun and generates immense amounts of power. Nikola Tesla also might have used this low energy neutrino radiation for his wireless transmission of electricity and in the magnifying transformer.

TESLA EXPERIMENTS

FIGURE 106: NIKOLA TESLA'S IN COLORADO SPRINGS

Nikola Tesla was a very well-known scientist famous for discovering the alternating current and his well-known experiments in Colorado Springs during 1899 for the wireless transmission of electricity. Tesla used massive coils with spark gaps to transmit electricity without wires between two towers miles apart. During these experiments, Tesla was able to transmit electricity between a transmitter and a receiver and also amplify the power many times than the original.

Tesla concluded that neutrinos from the Sun are related to scalar waves and can travel faster than the speed of light. Neutrinos were also the reason for the increased power that occurred between a transmitter to the receiver (magnifying transformer).

WIRELESS TRANSMISSION OF ELECTRICITY

During his famous experiments, Tesla showed that it is possible to transmit electricity without wires through electrical scalar waves. The power was also amplified almost ten times the original amount; this might be due to the low energy neutrino radiation from the Sun. The transfer of electricity occurred when the transmitter and receiver were in resonance.

It is now possible to conduct Tesla's lab experiments using Prof. Konstantin Meyl's kit available at www.meyl.eu. Using this kit makes it possible to conduct many experiments with scalar waves and understand their behavior. People performing the experiments using this kit could observe magnifying power at the receiver, similar to Tesla's experiments.

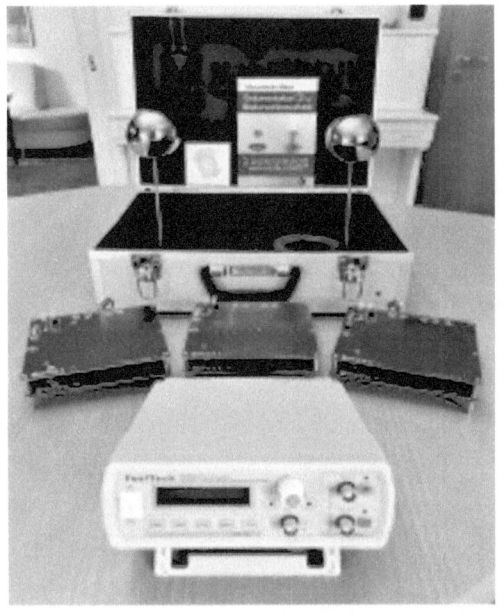

FIGURE 107: KONSTANTIN MEYL'S KIT FOR SCALAR WAVE EXPERIMENTS [45]

SCALAR WAVES

During his experimentation, Tesla also discovered scalar waves. Unlike EM waves, scalar waves travel faster than the speed of light and tap into neutrino radiation. This ability also makes them capable of passing through the Earth from one side to another without losing field strength. Additionally, unlike EM waves, scalar waves are longitudinal waves and were originally captured in Maxwell's equations [46].

All the waves generated by an antenna contain both longitudinal scalar waves and also transverse electromagnetic waves. For example, near a radio antenna, most of the waves will be scalar waves, and after some distance, they get converted to electromagnetic waves. We don't have a clear understanding of how the conversion happens between these two types of waves.

TYPES OF SCALAR WAVES

It was discovered by Prof. Konstantin Meyl that there are two main types of scalar waves:

1.) Electric Scalar Waves
2.) Magnetic Scalar Waves

Tesla used electric scalar waves in his experimentation with the wireless transmission of electricity. Meyl discovered the existence of magnetic scalar waves, and these are used by our human body and other biological systems, specifically for our cell communication [47].

BIOLOGICAL SYSTEMS

Prof. Konstantin Meyl discovered that magnetic scalar waves are used for cellular communication and can pass information between the DNA molecules. As stated previously, the only time electricity could pass without using a wire was when the transmitter and receiver were in resonance. And there are three unique requirements (as described by Tesla) that ensure this:

1.) The receiver and transmitter have to be at the same frequency.
2.) Opposite phase angle (180-degree phase change of the two waves) between the receiver and transmitter
3.) Same modulation between the receiver and the transmitter.

Interestingly, these three conditions are also unique to our body; hence our DNA doesn't pass information to other biological entities. It is theorized that so-called "junk DNA" is essential for creating this resonance [48] [49].

AURA AND DNA OSCILLATIONS

Benzene is an interesting molecule. It has a ring of carbon bonded with hydrogen, which shares electrons amongst each other. This free movement of electrons generates electromagnetic and scalar waves (as we know, moving charge in the perpendicular directions generates magnetic and electric waves, and accelerating charge creates an electromagnetic field). Within our body, all the biological molecules contain many similar ring structures. All these ring structures with accelerating and decelerating electrons generate electromagnetic and scalar waves used for cell communication. These waves also create the aura of our body. Hypothetically, reading this aura would allow us to continually cure any disease and maintain perfect health by merely correcting the specific waves within our bodies.

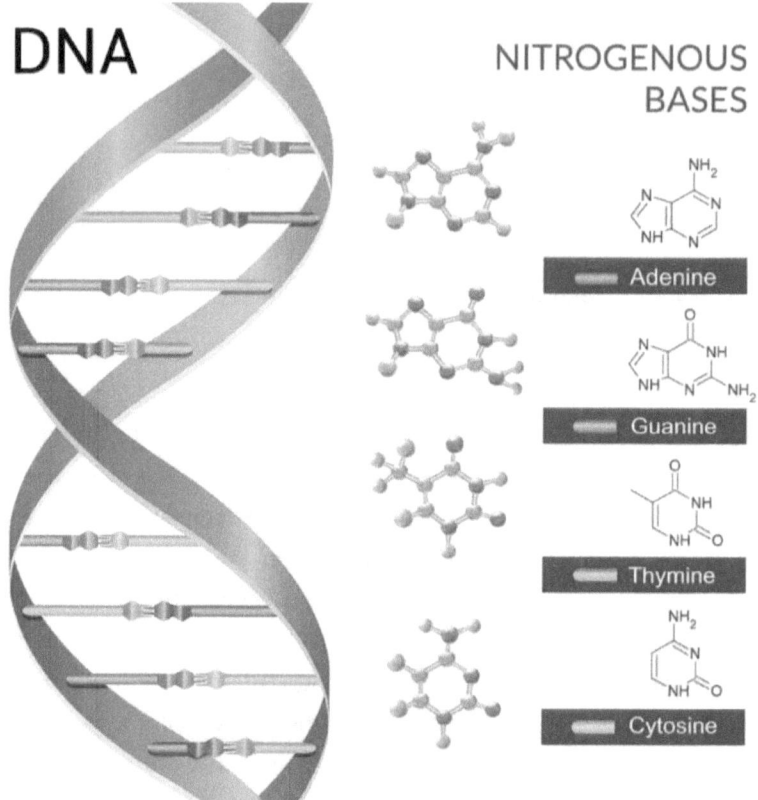

FIGURE 108: DNA MOLECULE AND THE RING STRUCTURES [50]

In our body, all biological molecules contain Benzene like ring structures (a DNA molecule structure has many ring structures). All these structures with accelerating and decelerating electrons generate electromagnetic waves and scalar waves for inter-cellular communication. These generated waves, in turn, create an aura for our body. If we can read this aura, we can cure any disease by sending correctional waves to the body. Maybe in the future, medicine will involve only correcting the aura. We are slowly progressing towards this goal; in fact, the Oberon biofeedback NLS system, one such system, tries to read the human aura and cures diseases [51].

INTERESTING CONJECTURES

For several thousands of years, humans have commonly used several spiritual practices within religion and medicine. Some of these practices are still deeply rooted within our culture to this day. Interestingly some of these ideas may originate in science and physics (though they are unscientific in the modern era). This section will go over some such practices and ideas while using physics to explain their relevance.

INFORMATION, MATTER, AND ENERGY

The Big Bang has been considered the start of the Universe. It consisted of a giant explosion of matter and energy, which slowly condensed into the Universe that we know and love today. Some Russian scientists have also speculated an atomic particle that carries information called 'Informion'. It may be possible that within our Universe, the total information, mass, and energy are considered constant (Information + Energy + Mass = Constant), and everything started based on an explosion of information. In this circumstance, the big bang was an "information bang" instead of where information particles were converted into matter and energy. This theory is highly speculative, and there is no proof. However, it does provide a different way of thinking about information within the Universe itself. The Russian scientists who discovered the Oberon biofeedback NLS system speculated this theory.

ASTROLOGY

Consistently, people (me included) have been surprised by the accuracy of some professional Indian Vedic astrological predictions. A large slew of people always tune in regularly to determine their future based on the stars. The concept of astrologically predicting the future remained unscientific and spiritual for a large part of history and even is today. Here is an interesting incite on astrology.

In our solar system, all the planets and the Sun are in resonance with each other. The frequency of this resonance is continuously changing as the planets are moving. Even the DNA within our cell resonates with these celestial bodies, and this process could be needed to exchange information and energy from the universe to ourselves (note all living beings consume energy from food and also from the universe in the form of radiation).

The best way to understand this is by analyzing a baby inside a mother's womb. During the time of pregnancy, when the baby is inside the mother's womb, it might be using the mother's calibrated resonance frequencies for information and energy exchange with the universe or solar system. But, when the baby is born, the baby's DNA needs to be recalibrated with all the new resonances and frequencies within the universe since it is now disconnected from the mother. These new calibrated frequencies will be unique to the baby depending on the time and place the baby is born (similar to a person's horoscope, which is just the planetary positions within the solar system at the time and place of the child's birth). These frequencies are integral for the brain and body's function, thereby affecting the child for the rest of their life.

These DNA calibration frequencies at the particular birth time and birthplace will be unique to the baby. These frequencies might determine the functioning of the baby's body and mind. As we know, a person's horoscope is nothing but planetary positions of the planets and Sun at the time and place of that person's birth. Hence, the positions of these stars and planets at the baby's time of birth are directly related to the calibrated DNA frequencies of the baby. This might be the reason these planetary positions dictate the baby's behavior and personality.

Again, this theory is nowhere close to being proven. It is speculation, to say the least. But, it might provide a bit of insight into how connected we truly are with the universe.

ECLIPSES AND ASTROLOGY

Solar eclipses and lunar eclipses are considered significant events in astrology. Astrologers predict major changes happening after these eclipses. The reason may be the following. During the eclipses, there will be a major change in the neutrino radiation on Earth as the moon will focus (like a lens) some of the slow-moving neutrinos to a specific location. This changed density of neutrinos might affect humans and the Earth, sometimes causing earthquakes and floods.

SUPERNOVA EFFECTS AND ASTROLOGY

In the past, during the dark ages, when there was intense violence, there was a supernova explosion in the Orion constellation. It was recorded there were two suns in the sky that morning for about a week. Maybe during the supernova explosion, neutrino radiation might have doubled and could affect our thinking and actions. The effects of these neutrinos might have made us more violent.

PRANA

In India, for thousands of years, prana (it is difficult to explain the concept of Prana with the known technical words) is considered the life-force or life-energy of all living beings. India is not the only culture that describes a form of life-energy specific to all living things. China also depicts something similar, and it is known as "chi," which is responsible for our survival and well-being. It is possible that ancient people were aware of neutrino radiation from the Sun and the universe. Neutrino radiation has been affecting all life on Earth, and our bodies might be absorbing different frequencies of this neutrino radiation based on our health, thoughts, and actions. This is another possible connection between ancient ideas and modern experimentation; obviously, nothing is truly proven. However, this provides yet another perspective to view the universe.

BIBLIOGRAPHY

[1] Wikipedia, "International System of Units," 27 11 2020. [Online]. Available: https://en.wikipedia.org/wiki/International_System_of_Units.

[2] CERN, "The Brout-Englert-Higgs mechanism," CERN, 2020. [Online]. Available: https://home.cern/science/physics/higgs-boson.

[3] A. Gupta, R. Resnick, D. Halliday and J. Walker, Physics for JEE (Main & Advanced) Vloume 1, New Delhi: Wiley , 2020.

[4] Wikipedia, "Ellipse," Wikipedia, 2020. [Online]. Available: https://en.wikipedia.org/wiki/Ellipse.

[5] muse, "A Deep Dive Into Brainwaves: Brainwave Frequencies Explained," muse, 2020. [Online]. Available: https://choosemuse.com/blog/a-deep-dive-into-brainwaves-brainwave-frequencies-explained-2/.

[6] M. K. Singhal, R. Resnick, D. Halliday and J. Walker, Physics for JEE (Main & Advanced) Volume 2, New Delhi: Wiley, 2019.

[7] Wikipedia, "Engine," 2020. [Online]. Available: https://en.wikipedia.org/wiki/Engine.

[8] Wikipedia, "Water turbine," 2020. [Online]. Available: https://en.wikipedia.org/wiki/Water_turbine.

[9] Wikipedia, "Single-phase generator," 2020. [Online]. Available: https://en.wikipedia.org/wiki/Single-phase_generator.

[10] U. Editors, "What is spacetime?," 2020. [Online]. Available: https://www.unrevealedfiles.com/en/what-is-spacetime/.

[11] Wikipedia, "Double-slit experiment," 2020. [Online]. Available: https://en.wikipedia.org/wiki/Double-slit_experiment.

[12] Wikipedia, "Periodic table," 2020. [Online]. Available: https://en.wikipedia.org/wiki/Periodic_table.

[13] G. Sparrow, The Planets, London: Quercus, 2006.

[14] NASA, "NASA," NASA, 2020. [Online]. Available: https://www.nasa.gov/topics/solarsystem/index.html.

[15] A. A. o. t. Universe, "Stars within 50 light years," 2020. [Online]. Available: http://www.atlasoftheuniverse.com/50lys.html.

[16] S. Hawking, A Brief History of Time, New York: Bantam Books, 1996.

[17] X. Qicheng and L. Jingyuan, The Earth's Crust, China: China Science and Technology Press, 1986.

[18] N. Geographic, "Earth's Interior," National Geographic, 2020. [Online]. Available: https://www.nationalgeographic.org/media/earths-interior/.

[19] A. Z. Jones and D. Robbins, String Theory for Dummies, Hoboken: Wiley, 2010.

[20] Xantox, "The Sun seen through the Earth in "neutrino light"," 2007. [Online]. Available: http://strangepaths.com/the-sun-seen-through-the-earth-in-neutrino-light/2007/01/06/en/.

[21] K. Meyl, "Experimental Kit," 2020. [Online]. Available: https://www.k-meyl.de/shop/product_info.php?products_id=7.

[22] K. Meyl, Scalar Waves, Villingen-Schwenningen: INDEL GmbH, 2003.

[23] K. Meyl, Documentation 1 on Scalar Wave Technology, Villingen-Schwenningen: INDEL GmbH, 2014.

[24] K. Meyl, DNA and Cell Resonance, Villingen-Schwenningen: INDEL GmbH, 2011.

[25] K. Meyl, Documentation 2 on Scalar Wave Medicine, Villingen-Schwenningen: INDEL GmbH, 2014.

[26] M. N. Today, "What is DNA and how does it work?," 2020. [Online]. Available: https://www.medicalnewstoday.com/articles/319818.

[27] O. Biofeedback, "Oberon Biofeedback," Oberon Biofeedback, 2020. [Online]. Available: https://oberondiagnostic.com/.

[28] Wikipedia, "List of Solar System objects by size," Wikipedia, 2020. [Online]. Available: https://en.wikipedia.org/wiki/List_of_Solar_System_objects_by_size.

[29] Wikipedia, "Circular motion," Wikipedia, [Online]. Available: https://en.wikipedia.org/wiki/Circular_motion.

[30] o. CNX, "Conservation of Angular Momentum," openstax CNX, 2020. [Online]. Available: https://cnx.org/contents/pZH6GMP0@1.409:JmOmbgy5@3/Conservation-of-Angular-Momentum.

[31] Wikipedia, "Inverse-square law," Wikipedia, 2020. [Online]. Available: https://en.wikipedia.org/wiki/Inverse-square_law.

[32] A.-L. Physics, "Longitudinal & Transverse Waves," A-Level Physics, 2020. [Online]. Available: https://alevelphysics.co.uk/notes/longitudinal-transverse-waves/.

[33] B. W. AOR, "HARMONIC VIBRATIONS," Odinic Rite, 2020. [Online]. Available: https://odinic-rite.org/main/harmonic-vibrations/.

[34] Tsunami, "Tsunami," Tsunami, 2020. [Online]. Available: https://www.wikiwand.com/en/Tsunami.

[35] Wikipedia, "Stress–strain curve," Wikipedia, 2020. [Online]. Available: https://en.wikipedia.org/wiki/Stress%E2%80%93strain_curve.

[36] Pinclipart, "Elastic Potential Energy," Pinclipart, 2020. [Online]. Available: https://www.pinclipart.com/pindetail/iRohimw_energy-clipart-elastic-potential-energy-elastic-energy-examples/.

[37] K. Chaudhari, "Poisson's Ratio," Engineer Gallery, 2020. [Online]. Available: http://www.engineersgallery.com/poissons-ratio/.

[38] Wikipedia, "Harmonic," Wikipedia, 2020. [Online]. Available: https://en.wikipedia.org/wiki/Harmonic.

[39] T. D. o. P. a. Astronomy, "The University of Western Ontario," 2020. [Online]. Available: HTTP://AQUARID.PHYSICS.UWO.CA/IMAGES/NEW/SOURCE%20OF%20INFRASOUND.JPG.

[40] Wikipedia, "Ultrasound," 2020. [Online]. Available: https://en.wikipedia.org/wiki/Ultrasound.

[41] v. S. V, "Why do some seismograph stations receive both primary and secondary waves from an earthquake but other stations don't?," 2020. [Online]. Available: https://socratic.org/questions/why-do-some-seismograph-stations-receive-both-primary-and-secondary-waves-from-a.

[42] B. Greene, "What the hell are gravitational waves?," 2020. [Online]. Available: https://rebelscumradio.com/2016/03/10/hell-gravitational-waves-dr-brian-greene-explains/.

[43] Infofavour, "What we know is just handful & what we do not know is the remaining of Earth..," 2020. [Online]. Available:

https://infofavour.blogspot.com/2016/10/first-second-and-third-law-of.html.

[44] A. Samudra, "Stefan Boltzmann Law," 2020. [Online]. Available: https://atomstalk.com/blogs/stefan-boltzmann-law/.

[45] R. Nave, "Radiation Curves," 2020. [Online]. Available: http://hyperphysics.phy-astr.gsu.edu/hbase/wien.html.

[46] Unknown, "Interpretation of Pressure on Kinetic Theory of Gases," 2020. [Online]. Available: https://simplexitysolutions.blogspot.com/2015/09/interpretation-of-pressure-on-kinetic.html.

[47] Wikipedia, "Capacitor," 2020. [Online]. Available: https://en.wikipedia.org/wiki/Capacitor.

[48] M. Irving, "Earth's magnetic field is weakening – but it's not about to reverse," 2018. [Online]. Available: https://newatlas.com/magnetic-field-reversal-not-imminent/54426/.

[49] Wikipedia, "File:Right hand rule cross product F=J×B.svg," 2020. [Online]. Available: https://en.wikipedia.org/wiki/File:Right_hand_rule_cross_product_F%3DJ%C3%97B.svg.

[50] Wikipedia, "Solenoid," 2020. [Online]. Available: https://en.wikipedia.org/wiki/Solenoid.

[51] Wikipedia, "Amplitude modulation," 2020. [Online]. Available: https://en.wikipedia.org/wiki/Amplitude_modulation.

INDEX

A

acceleration, 3, 5, 9, 10, 14, 15, 19, 20, 23, 28, 29, 36, 37, 39, 44, 159, 163
Acceleration, 9, 19, 37
acceleration due to gravity, 15, 19, 23, 163
action-at-a-distance, 17
action-reaction, 15
aerospace, 28
air pressure, 48
air resistance, 14, 17, 22, 23, 27, 28, 46
Albert Einstein, 30, 159, 170, 174
Alpha Centauri, 202, 203
alpha particle, 167
Alternating currents, 125
Amedeo Avogadro, 106
ampere, 3
amplitude, 55, 56, 62, 63, 65, 66, 78, 84, 85, 137, 138, 144, 145
Amplitude modulation, 137
Amplitude Modulation, 61, 137
Andromeda galaxy, 206, 208
Angular, 11, 35, 37
angular momentum, 11, 29, 35, 36
angular velocity, 35
Antennas, 138
antinode, 65
Archimedes' Principle, 50
asteroids, 30, 189, 194, 195, 197
astrology, 189, 235, 236
astronomical bodies, 63, 64, 223
astronomical object, 11
astronomy, viii, 137, 189

atmosphere, 48, 59, 97, 106, 113, 115, 141, 191, 192, 193, 194, 195, 197, 198, 199, 212, 214, 215, 216, 217, 218
atmospheric pressure, 48, 157
atom, 2, 11, 108, 112, 122, 140, 146, 147, 157, 167, 168, 176, 177, 178, 179, 181, 182, 185, 213, 227
atomic number, 180, 181
atomic particles, 54, 56, 75, 112, 113, 147, 160, 167, 169, 170, 182, 220, 221, 222, 228
atomic symbol, 181
atomic weight, 182
atoms, 47, 52, 70, 75, 93, 95, 102, 106, 107, 108, 112, 122, 136, 145, 146, 147, 155, 166, 173, 176, 177, 178, 179, 181, 183, 185, 186, 213, 214, 215, 216, 217, 227, 228
Atoms, 47, 112, 147, 176, 178, 179
Avogadro's number, 106, 107

B

battery, 115, 118, 123, 132, 146, 147, 149
beats, 76, 78, 79
Benzene, 233, 234
Bernoulli's equation, 52, 53
Bernoulli's principle, 52, 53
Beta Geminorum, 203
Big Bang, 223, 235
Binaural Beats, 79
black body, 166, 167
black holes, 88, 89, 165, 205, 206, 209, 210, 221, 228
blueshift, 64, 164
Boltzmann, 98, 99, 108
brain, x, 76, 79, 83, 84, 121, 228, 236

brainwave, 79, 80, 83
brainwaves, 83, 84
buckminsterfullerene, 186
buoyant force, 50, 51

C

calculus, 4, 127, 170
Canis Major, 207, 208
capacitance, 112, 117, 124
Capacitance, 117
cell phones, 118
Celsius, 91, 94
center of gravity, 22
center of mass, 21
centrifugal force, 38, 69
centripetal force, 38
Centripetal force, 38
CGS system, 3
charge, 109, 110, 112, 113, 114, 115, 116, 117, 118, 122, 124, 125, 128, 129, 131, 132, 147, 155, 175, 177, 182, 216, 228, 233
chemical bonds, 71, 136, 146
chemical energy, 146, 147
circle, 9, 11, 30, 31, 34, 38, 69, 133, 172, 205, 206
circuit, 117, 118, 124, 125, 153, 215
circuits, 118, 124, 125, 139
circular, 8, 9, 29, 31, 34, 35, 38, 121, 133, 172, 178, 185, 187, 205
circular motion, 29, 34, 35
circular rotation, 9
classical mechanics, 5, 13, 22, 166
Coherence, 145
collision, 18, 19, 88, 89, 105, 196, 207, 208
collisions, 11, 18, 19, 104, 105, 172, 176, 184, 206, 208, 217
Combustion Stroke, 154
components of a vector, 4
compressibility, 51, 52

compression, 52, 66, 82
Compression Stroke, 154
computers, 118, 172, 174
Conductors, 115
conservation of energy, 43, 45
conservation of mechanical energy, 46
conservation of momentum, 18
Constructive interference, 78, 145
Copernicus, 32
cosmic rays, 180, 227, 228
cosmic waves, 88
cosmically charged particles, 41
cross-sectional area, 27
crust, 13, 190, 191, 192, 194, 218
current, 3, 112, 116, 117, 118, 122, 123, 124, 125, 131, 132, 133, 146, 147, 148, 149, 150, 153, 155, 194, 216, 217, 229, 230

D

dark matter, 1, 187, 204, 208, 224
density, 21, 39, 48, 85, 105, 109, 119, 199, 210, 212, 214, 216, 237
Density, 48
Deoxyribonucleic Acid, 75
destructive interference, 78, 145
diffraction, 61, 62, 144, 145
Diffraction, 61, 62, 144
Direct current, 125
direction, 3, 4, 5, 6, 7, 8, 9, 10, 15, 16, 19, 20, 23, 34, 35, 36, 37, 43, 48, 50, 57, 58, 59, 60, 61, 64, 73, 82, 109, 114, 116, 119, 121, 122, 123, 125, 132, 140, 142, 143, 146, 150, 155, 161, 197, 207, 210
displacement, 5, 7, 8, 9, 34, 35, 46, 54, 57, 99, 143, 159
Distance, 8, 17
DNA, 75, 232, 233, 234, 235, 236
Doppler Effect, 64, 156, 223
double-slit, 166, 167, 168, 171

drag, 27, 217
dynamic torque, 36
Dynamics, 16

E

Earth, 5, 13, 15, 18, 19, 20, 21, 28, 29, 31, 40, 41, 48, 64, 68, 69, 81, 82, 95, 97, 98, 111, 113, 115, 119, 133, 137, 139, 141, 159, 160, 161, 164, 183, 187, 189, 190, 191, 192, 193, 194, 195, 196, 198, 199, 200, 202, 203, 204, 206, 208, 209, 212, 213, 214, 215, 216, 217, 218, 219, 220, 223, 227, 228, 229, 231, 236, 237
Earth's magnetic field, 217
Earth's core, 20
earthquake, 82
eccentricity, 34
eclipses, 236
ecliptic plane, 34
Einstein, 56, 88, 89, 162, 163, 174, 175, 209, 220, 221
elastic collisions, 19
elasticity, 70, 71
Elasticity, 70
electric charge, 112, 113, 115, 124, 132
electric field, 4, 110, 112, 113, 114, 115, 120, 126, 128, 129, 130, 131, 132, 133, 150, 155, 225
electric field lines, 113, 114, 132
electric potential, 112, 115, 116
electrical force, 17
electricity, 3, 110, 112, 115, 118, 121, 126, 131, 132, 138, 147, 148, 149, 151, 153, 216, 220, 229, 230, 232
Electricity, 112, 230
Electromagnetic radiation, 136
electromagnetic waves, 54, 63, 134, 135, 137, 232
Electromagnetic Waves, 56, 133
Electromagnetism, 121, 225

electromotive force, 124
electron, 11, 113, 114, 121, 122, 126, 140, 146, 167, 168, 169, 170, 171, 177, 178, 181, 182, 183, 184, 185, 215, 221, 227
electrons, 5, 56, 112, 113, 115, 116, 118, 121, 122, 136, 138, 140, 146, 147, 152, 153, 155, 166, 167, 168, 172, 174, 176, 177, 178, 179, 181, 183, 185, 214, 216, 227, 233, 234
Electrons, 112, 177, 178
elementary particles, 1, 176
ellipse, 30, 31, 33, 34
elliptical orbit, 31, 191
elliptical orbits, 30, 31, 224
energy, 16, 18, 19, 38, 39, 42, 43, 44, 45, 46, 53, 54, 55, 65, 66, 71, 72, 75, 85, 86, 90, 92, 93, 94, 95, 97, 98, 99, 101, 102, 103, 104, 107, 108, 109, 115, 117, 120, 123, 124, 125, 126, 135, 136, 137, 138, 139, 140, 145, 146, 147, 148, 151, 152, 153, 154, 155, 157, 158, 167, 170, 172, 174, 175, 176, 177, 179, 181, 182, 184, 187, 205, 214, 216, 222, 226, 227, 229, 231, 235, 236, 237
Energy, 42, 43, 45, 65, 72, 92, 107, 120, 146, 147, 226, 229, 235
entropy, 101, 102, 103, 104, 222
Entropy, 101, 102, 104
equilibrium, 37, 39, 66, 91, 92, 94, 107
Equilibrium, 39
equipartition, 107
Exhaust Stroke, 154
exosphere, 40, 191, 217
expansion of solids, 95
external force, 13, 14, 15

F

Faraday's Law, 127, 131, 132, 133
field flux, 109, 128, 130
field lines, 109, 111, 113, 114, 119, 128, 130, 132, 195

fields, 109, 110, 111, 114, 118, 119, 120, 121, 123, 126, 128, 129, 130, 131, 132, 133, 134, 173, 180, 184, 189, 190, 222, 226
first law, 14, 31, 37, 39, 128
fission, 141, 147, 179, 186, 187
fluid, 27, 49, 50, 51, 52
flux, 112, 114, 124, 128, 129, 130, 132
foci, 30, 34
force, 5, 6, 10, 13, 14, 15, 16, 17, 20, 21, 22, 25, 26, 27, 28, 29, 31, 36, 37, 38, 39, 43, 44, 45, 47, 48, 50, 51, 66, 69, 70, 71, 72, 73, 74, 75, 93, 109, 110, 114, 118, 121, 122, 124, 132, 133, 147, 150, 152, 153, 155, 158, 177, 178, 180, 184, 187, 217, 220, 222, 224, 237
force of gravity, 16, 22
forces, 5, 6, 11, 13, 14, 15, 16, 17, 25, 26, 29, 36, 38, 39, 43, 45, 47, 69, 70, 74, 75, 124, 126, 185, 187, 199
FPS system, 3
free motion, 5
frequency, 56, 62, 63, 64, 65, 66, 67, 76, 77, 78, 79, 80, 83, 84, 87, 126, 137, 138, 139, 156, 157, 167, 174, 197, 216, 221, 223, 232, 235
Frequency modulation, 138
Frequency Modulation, 61, 138
friction, 17, 25, 26, 27, 46, 140
fuel, 16, 24, 27, 125, 147, 153, 154, 155, 208, 210
fundamental units, 3, 38
fusion, 141, 147, 157, 158, 190

G

galaxies, xi, 29, 164, 165, 187, 188, 189, 204, 205, 206, 207, 208, 220, 223, 224
Galaxies, 204, 205, 206, 224
Galileo Galilei, 16
gamma rays, 126, 141
gamma synchrony, 83

Gamma-rays, 141
gases, 47, 52, 70, 76, 94, 104, 105, 146, 178, 181, 194, 195, 199, 216
Gauss' law, 130
Gauss's Law for Electric Fields, 128, 129
generator, 118, 123, 132, 152
geostationary orbit, 40
gluons, 166, 184
gravitational attraction, 29, 162
gravitational constant, 29
gravitational field, 5, 43, 111, 164, 195, 198, 209
gravitational fields, 111, 226
gravitational force, 6, 17, 21, 29, 38, 44
gravitational interaction, 5
Gravitational lensing, 165
gravitational potential energy, 42
gravitational pull, 5, 13, 20, 28, 69, 163, 187, 189, 195, 197, 200, 207, 209, 212, 224
gravitational waves, 54, 89
gravity, 5, 14, 15, 16, 19, 20, 22, 23, 26, 28, 29, 33, 43, 44, 45, 48, 88, 162, 163, 164, 187, 190, 194, 202, 204, 205, 209, 210, 217, 222, 224

H

Hall Effect, 121, 122
harbor resonance, 67
harmonic, 54, 65, 66
harmonics, 76, 78
heat, 2, 39, 45, 46, 54, 85, 90, 93, 94, 97, 98, 99, 103, 105, 108, 109, 134, 136, 140, 147, 148, 157, 167, 186, 187, 191, 192, 194, 196
Heisenberg, 168, 169, 177, 226
Helmholtz resonance, 66, 67
Higgs Boson, 1, 6, 182, 184, 185
Higgs field, 6, 184
high school, viii, 1, 142
high-pitch, 76

Hooke's law, 71
horizontal, 22, 23, 24, 52, 53, 161
horizontal component, 22
horizontally, 22, 41, 57, 161
horsepower, 38
human brain, 76
Huygens's principle, 142

I

Incompressible, 53
Inductance, 124
inductor, 120, 124
inelastic, 18, 19
inertia, 6, 11, 25, 35, 37, 66, 69, 162, 187
Inertia, 6, 7, 14, 37
influence of gravity, 19
Infrared, 136
infrasound, 80
inner core, 120, 212, 219
insulators, 115, 129
Intake Stroke, 154
interference, 61, 78, 144, 145, 156, 167, 171
Interference, 61, 144
Internal Combustion engines, 153
internal force, 13
International System of Units, 3
interstellar, 15, 56
inverse square law, 55, 225
ionosphere, 139, 214, 215, 216
Isaac Newton, 142
isolated systems, 45

J

jet engine, 24, 154, 155
Jupiter, 31, 194, 195, 196, 198, 200, 201, 203

K

kelvin, 3
Kepler, 1, 30, 31, 32, 224
Kepler's laws, 30
Kepler's Laws, 1, 30, 224
kilogram, 3
kinematics, 5, 36, 37
Kinematics, iii, 5, 162
kinetic energy, 19, 42, 43, 44, 45, 65, 90, 97, 105, 108, 151
Kinetic energy, 42
kinetic friction, 25, 26, 27
kinetic theory, 104, 105
Konstantin Meyl, 222, 225, 231, 232
Kuiper belt, 201

L

laser, 145, 146
Law of Conservation, 12, 17, 35, 92
Law of Cooling, 100
laws of motion, 13, 160
length contraction, 159, 161, 162
Lenz's law, 132
Lift, 27
lift induced, 27
light, 2, 3, 11, 45, 59, 60, 62, 64, 89, 97, 98, 99, 100, 109, 118, 124, 126, 127, 133, 134, 135, 136, 139, 142, 143, 144, 145, 146, 147, 155, 157, 159, 160, 161, 163, 164, 165, 166, 167, 168, 169, 171, 174, 175, 182, 187, 190, 201, 202, 203, 204, 205, 206, 207, 208, 209, 210, 213, 216, 223, 225, 227, 228, 230, 231
lightening, 80
lightning, 85, 86, 173
linear velocity, 35
liquid, 24, 27, 47, 50, 52, 68, 76, 97, 102, 157, 158, 192, 193, 195, 196, 197, 202, 217, 219
longitudinal, 56, 73, 75, 77, 82, 225, 232

longitudinal waves, 56, 77, 82, 225
Lorentz, 121, 122
Lorentz Force, 121, 122
Low Earth Orbit, 41
low-pitch, 76
luminous intensity, 3
lunar tides, 69

M

Mach, 27
macroscopic objects, 13
magnetic dipole, 122, 180
magnetic field, 4, 41, 109, 110, 111, 118, 119, 120, 121, 122, 123, 124, 126, 130, 131, 132, 133, 149, 153, 190, 192, 193, 194, 195, 197, 198, 217, 218, 219, 224, 225, 227
magnetic fields, 110, 133
magnetic force, 17
magnetic pole, 120
magnetism, 110, 118, 121, 126, 130, 131, 132, 149, 220
magnetized, 19, 119, 194
magnets, 119, 132, 224
magnitude, 3, 4, 6, 8, 9, 16, 19, 20, 43, 47, 48, 50, 119, 167
major axis, 34
mantle, 192, 199, 218, 219
Mars, 6, 186, 190, 193, 194, 200, 201, 218, 227, 228
mass, 2, 3, 5, 6, 10, 11, 14, 17, 18, 19, 21, 28, 29, 37, 38, 43, 48, 50, 56, 66, 133, 157, 162, 163, 175, 177, 179, 180, 181, 183, 184, 185, 187, 190, 194, 200, 203, 204, 206, 209, 210, 212, 223, 228, 235
Mass, 5, 6, 21, 235
massive, xi, 10, 11, 29, 69, 88, 133, 153, 162, 163, 164, 165, 174, 184, 194, 195, 202, 203, 208, 209, 220, 228, 230
mathematical formula, 25, 220

matter, 3, 6, 28, 48, 54, 57, 58, 75, 93, 119, 126, 134, 163, 176, 182, 184, 187, 208, 221, 222, 223, 228, 235
Max Planck, 167
maximum height, 22, 23
Maxwell, 108, 126, 127, 128, 129, 131, 132, 220, 224, 225, 231
measurement, 2, 42, 90, 130, 169, 170, 226
measurements, 1, 2, 8, 187, 214
mechanical wave, 65, 66
Mechanical Waves, 56
mechanics, viii, x, 13, 16, 151, 166, 168, 220
medium Earth orbit, 41
Mercury, 191
mesosphere, 213, 214
metaphysical, 166
metric system, 3
microwave oven, 139, 153
Microwaves, 139
Milky Way, 189, 204, 205, 206, 207, 208, 224
minor axis, 34
modulating the waves, 61
mole, 3, 106, 107
molecules, 27, 52, 63, 70, 75, 86, 93, 96, 102, 104, 105, 106, 107, 140, 153, 158, 172, 176, 179, 185, 186, 213, 214, 216, 217, 232, 233, 234
momentum, 4, 10, 11, 16, 17, 19, 29, 35, 36, 133, 169, 170, 182
Momentum, 4, 10, 11, 12, 17, 35
moon, 19, 68, 69, 200, 236
Moon, 6, 189, 190, 191, 199, 200
moons, 30, 189, 194, 195, 196, 197, 199, 200
motion, 4, 5, 6, 7, 8, 10, 11, 13, 14, 15, 16, 22, 23, 24, 25, 26, 27, 29, 30, 32, 33, 34, 35, 36, 39, 40, 42, 43, 54, 64, 66, 93, 104, 121, 123, 142, 154, 159,

160, 162, 164, 173, 175, 199, 205, 220, 224
motors, 121, 150
moving objects, 4, 160
Mt. Everest, 20, 45

N

natural frequency, 62
nature, x, 15, 56, 57, 62, 63, 84, 87, 109, 110, 167, 169, 174, 180, 203
Neil Bohr, 167
Neptune, 197, 198, 199, 200, 201
net force, 13, 15, 20
neutrino, 56, 180, 182, 183, 228, 229, 231, 236, 237
Neutrino, 56, 183, 228, 229, 237
neutrinos, 1, 182, 183, 228, 229, 230, 236, 237
Neutrinos, 182, 228, 230
neutron, 157, 158, 179, 180
neutrons, 5, 56, 147, 158, 176, 177, 178, 179, 180, 181, 183, 184
Newton, viii, xi, 1, 5, 6, 13, 14, 15, 16, 24, 26, 29, 37, 39, 100, 124, 135, 142, 159, 160, 162, 187, 220
Newton's Laws, 1, 13, 29
Newtonian physics, xi, 13
Newtons, 5
Nikola Tesla, 130, 229, 230
nonconductors, 115, 129
normal force, 16, 26, 27, 38
Nuclear energy, 147
nucleus, 122, 146, 147, 167, 177, 178, 179, 181, 182, 227

O

octave, 77
Ohm's Law, 116
one-dimension, 7
optical fibers, 60

Optics, 142
orbiting, 15, 36, 164, 177, 189
orbits, 30, 34, 69, 147, 164, 167, 178, 198, 200, 204
oscillations, 54, 56, 57, 58, 93
outer core, 217, 219
overtones, 78
ozone layer, 213

P

parabola, 22
parasitic, 27
parasitic drag, 27
particle colliders, 6, 187
Pascal's Law, 49
period of oscillation, 56
periodic oscillations, 54
periodic table, 180, 181, 182, 185
permittivity, 129
photon, 169, 171, 174, 175, 185, 221
photons, 1, 56, 126, 133, 135, 145, 146, 155, 166, 168, 171, 172, 174, 175, 209, 216
physical phenomenon, 18
pitch, 77, 156, 169
planet, 11, 20, 28, 31, 32, 36, 115, 161, 187, 189, 191, 192, 194, 195, 196, 197, 199, 200, 202, 203, 204, 209, 212, 213, 217
Planet X, 200
Planetary Motion, 1, 33
planets, xi, 29, 30, 31, 32, 33, 36, 75, 164, 177, 189, 190, 191, 192, 193, 197, 198, 199, 200, 204, 206, 209, 217, 220, 223, 224, 235, 236
plasma, 47
Pluto, 198, 199, 200
Poisson's ratio, 73, 74
potential energy, 42, 43, 44, 45, 53, 66, 72, 114
Potential energy, 42

potential to do work, 42
Power, 38, 42, 65, 118
prana, 237
pressure, 47, 48, 49, 51, 52, 53, 66, 68, 77, 85, 86, 102, 105, 106, 126, 133, 158, 186, 195, 196, 198, 212, 216, 219
Pressure, 47, 133
projectile, 22, 23
propulsion, 186
protons, 5, 56, 121, 147, 176, 177, 178, 179, 180, 181, 182, 183, 184, 227
Proxima Centauri, 203
p-waves, 82
P-waves, 82

Q

quantum bits, 172
quantum computer, 173, 174
quantum entanglement, 172, 173, 221
quantum field theory, 1
quantum level, 29, 166
quantum mechanics, viii, 1, 166, 168, 169, 170, 171, 176, 220, 221, 226
quarks, 166, 176, 177, 178, 179, 180, 182, 183, 184
qubits, 172, 173, 174

R

Radar, 139, 155, 156
radars, 118
radiation, 55, 56, 94, 98, 99, 100, 126, 127, 133, 135, 136, 137, 138, 140, 141, 166, 180, 197, 213, 214, 215, 216, 217, 227, 228, 229, 231, 236, 237
Radiation, 41, 95, 97, 126, 133, 146, 166, 227, 228
radio, 54, 56, 88, 89, 126, 137, 138, 139, 144, 153, 156, 215, 216, 232
radio transmissions, 56
radio waves, 126, 137, 138, 139, 215

radioactive, 41, 170, 180
rainbow, 59, 60, 135, 142, 144
reaction, 15, 16, 26, 146, 147, 148, 158, 179
redshift, 64, 164
reference point, 38, 45
reflection, 59, 60, 87, 142, 143, 173
Reflection, 59, 60, 142
refraction, 59, 60, 142, 143
relativity, 159, 162, 209
REM, 85
resonance, 62, 63, 64, 66, 67, 76, 215, 216, 231, 232, 235
Resonance, 62, 63, 66, 67, 215, 216
revolution, 32, 41
rocket, 16, 24
rocket engine, 24
Rockets, 24
rotation, 9, 33, 34, 38, 41, 191, 192, 193, 194, 196, 197, 198, 199, 217
rotational equilibrium, 37
rotational inertia, 35
roughness, 26
Rutherford, 141, 167, 168, 179

S

Sagittarius Dwarf Galaxy, 207, 208
satellite, 15, 40, 41, 190, 215, 217
satellites, 31, 40, 41, 164, 195, 207, 213, 216, 217, 218
Satellites, 40, 164
Saturn, 31, 194, 195, 196, 197
scalar waves, 54, 58, 63, 231, 232, 233, 234
scalars, 3
Scalars, 3
Schrödinger's cat, 170, 171
Schrödinger's wave equation, 170
Schumann, 215, 216
scientists, xi, 2, 6, 18, 88, 89, 104, 106, 140, 149, 155, 157, 158, 166, 171, 172,

174, 175, 176, 177, 178, 179, 183, 184, 185, 186, 198, 200, 206, 210, 213, 214, 217, 222, 226, 229, 235
Scientists, 5, 91, 107, 140, 177, 203, 208, 222, 223
second law, 14, 31, 37, 129
seiche, 67
Seiche, 67
Seismic, 76, 81
seismic waves, 56
Self-inductance, 124
semi-major axis, 32
shock wave, 85
shockwaves, 27
SI units, 3, 48
Solar panels, 155
solar system, 11, 30, 31, 34, 189, 190, 191, 195, 197, 198, 199, 200, 201, 203, 206, 207, 208, 227, 235, 236
solenoid, 123
solid, 24, 47, 70, 76, 82, 95, 191, 194, 219
sonar, 77, 87
sonic boom, 86
sonofusion, 157
Sonoluminescence, 157
sound, 17, 19, 27, 52, 54, 55, 56, 62, 63, 64, 66, 76, 77, 78, 79, 80, 85, 86, 93, 109, 126, 137, 138, 139, 156, 157, 162, 174, 210
sound waves, 54, 56, 63, 76, 77, 80, 86, 87, 137, 138, 156
Sound waves, 58, 86
sounds, 17, 42, 61, 64, 66, 76, 77, 80, 87, 107, 138, 166
space, 2, 5, 8, 11, 14, 15, 18, 24, 28, 30, 52, 54, 56, 76, 88, 89, 98, 99, 106, 113, 122, 126, 129, 133, 140, 159, 161, 162, 163, 164, 165, 169, 170, 175, 186, 187, 189, 197, 201, 206, 209, 216, 217, 218, 220, 223, 226, 227
space-time, 89, 163, 164

spectrum, 121, 134, 135, 136, 139, 141, 173
speed, 2, 3, 8, 9, 10, 11, 15, 18, 20, 22, 26, 27, 29, 31, 32, 35, 36, 38, 40, 43, 53, 56, 58, 59, 60, 68, 76, 85, 86, 129, 139, 155, 159, 160, 161, 163, 168, 169, 173, 174, 175, 177, 191, 194, 209, 210, 223, 224, 225, 227, 230, 231
speed of light, 159, 160, 161, 175, 209, 210, 223, 226
speed of sound, 76
spring's vibration, 58
standing wave, 57
star, 36, 64, 99, 133, 187, 190, 201, 202, 203, 204, 205, 206, 208, 210
stars, xi, 29, 64, 75, 88, 98, 133, 135, 164, 165, 187, 189, 190, 201, 202, 203, 204, 205, 206, 207, 208, 210, 220, 223, 224, 228, 235, 236
static friction, 25, 26, 27
static torque, 36
Strain, 71
stratosphere, 213
stress, 71, 74, 85, 185
string theory, viii, 1, 220, 221, 222
Subatomic, 166
subconscious, 76
subliminal, 76
Subliminal messaging, 76
submarines, 77, 139
subsonic, 76
Sun, 2, 13, 30, 31, 32, 34, 56, 68, 69, 95, 96, 97, 133, 136, 142, 159, 177, 183, 189, 190, 191, 192, 193, 194, 196, 197, 198, 199, 200, 201, 202, 203, 204, 206, 208, 210, 213, 215, 216, 217, 224, 228, 229, 230, 231, 235, 236, 237
superconductors, 116, 173, 186
supernova, 54, 88, 237
supersonic, 27, 76, 85, 86
supersonic fighter, 76
s-waves, 82

T

tangential velocity, 9
technology, x, 28, 48, 145, 149, 151, 186, 189, 199, 222
temperature, 2, 3, 39, 85, 90, 91, 92, 93, 95, 100, 101, 102, 103, 104, 105, 106, 115, 135, 136, 158, 166, 186, 192, 194, 212, 216
Temperature, 90, 92
tension, 17
Tesla, 130, 230, 231, 232
theory of relativity, 1, 56, 88, 159, 160, 161, 162, 175, 220, 221, 223
thermal energy, 19, 90, 103, 135, 154
thermodynamic equilibrium, 39
thermodynamics, 39, 91, 94, 101, 102, 107
Thermodynamics, iii, 1, 19, 90, 91, 92, 102, 104
thermometers, 97
thermosphere, 213, 216
third law, 6, 25, 32, 131
three-dimensions, 7
thrust, 15, 24
time, x, xi, 2, 3, 8, 9, 11, 14, 23, 26, 30, 31, 32, 34, 35, 39, 55, 56, 57, 59, 61, 65, 66, 68, 83, 84, 85, 87, 88, 89, 103, 106, 116, 125, 127, 135, 139, 144, 156, 158, 159, 160, 161, 162, 163, 164, 165, 166, 169, 170, 171, 173, 175, 186, 196, 199, 208, 210,214, 215, 216, 217, 220, 226, 227, 232, 236
time dilation, 159, 160, 161, 162
torque, 35, 36, 37, 38, 75
Torque, 36, 37, 38
total internal reflection, 60
transferring energy, 54
transverse, 56, 57, 82, 126, 225, 232
transverse waves, 57, 231
troposphere, 212, 214
tsunami, 68
tsunamis, 67, 68
Tsunamis, 68
turbines, 148, 151
Turbines, 151
two-dimensions, 7

U

ultrasonic, 76
ultrasound, 80, 157
Ultrasound, 80
Ultraviolet, 126, 136
unbalanced force, 22
Uncertainty Principle, 168, 169, 177, 226
underwater, 48, 68, 77, 87, 196
uniform disc, 21
universe, xi, 15, 29, 54, 70, 98, 109, 136, 140, 147, 170, 172, 175, 176, 177, 182, 184, 185, 187, 189, 203, 205, 206, 209, 210, 221, 222, 225, 226, 236, 237
Universe, xi, 1, 2, 6, 11, 29, 75, 88, 90, 127, 190, 220, 223, 224, 235
Uranus, 196, 197, 198, 199, 202

V

vacuum, xi, 24, 48, 76, 95, 126, 129, 133, 134, 144, 159, 163, 173, 174, 175, 210, 226
vacuum energy, 226
Van Allen Radiation Belts, 41
vector algebra, 4
vector quantity, 4, 8, 9, 10, 34, 37, 38, 43, 48
vectors, 3, 4, 16
Vectors, 3
velocity, 5, 6, 8, 9, 10, 11, 15, 18, 22, 23, 24, 27, 35, 36, 38, 39, 43, 46, 56, 57, 58, 59, 126, 159, 160, 209, 210
Venus, 192, 199, 212
vertical, 22, 23, 24, 44, 52, 161, 212
vertical component, 22

vertically, 22, 57
vibrations, 54, 63, 70, 75, 76, 77, 81, 109, 221
volume, 3, 48, 50, 52, 53, 55, 66, 74, 79, 95, 102, 107

W

water waves, 56, 57, 61, 65
Water waves, 55, 66
wave, 27, 54, 55, 56, 57, 58, 59, 60, 61, 62, 65, 66, 67, 68, 76, 78, 82, 85, 86, 87, 127, 133, 134, 137, 138, 142, 143, 144, 145, 153, 156, 157, 158, 166, 167, 168, 169, 170, 171, 172, 173, 174, 176, 177
wave propagation, 56, 59, 60, 82
wave's intensity, 55
wavelength, 56, 58, 61, 99, 126, 135, 141, 144, 145, 146, 169
wavelengths, 100, 126, 134, 135, 136, 139, 142, 169, 213
waves, 54, 55, 56, 57, 58, 59, 61, 62, 63, 64, 65, 66, 68, 75, 76, 77, 78, 79, 80, 81, 82, 83, 84, 85, 86, 87, 88, 89, 126, 127, 129, 133, 134, 136, 137, 138, 139, 140, 144, 145, 146, 153, 156, 158, 166, 167, 169, 170, 176, 214, 215, 216, 222, 225, 228, 230, 231, 232, 233, 234
weight, 5, 26, 28, 29, 50, 181
weightlessness, 28
WiFi, 56
wings, 27, 53
wireless communication, 54
world, 1, 7, 18, 19, 56, 85, 119, 138, 139, 142, 148, 149, 167, 170, 172, 174, 186, 202, 204, 212, 221, 222

X

X-rays, 126, 140, 141, 167

Z

zero-point energy, 90
Zeta Reticuli, 204

ABOUT THE AUTHOR

Anisha Yeddanapudi is a graduating senior in Mountain House High School, California. She has always been interested in physics and research. The combination of these two passions led to the publication of this book. When she isn't spending time reading or learning, you can always find her painting or enjoying long walks on the beach. Physics for All is Anisha's first book on the topic and extensively covers the subject she loves.

This book took several months to a year to develop and was edited and proofread by Krishna Yeddanapudi to ensure consistency.

Author's contact details:

Email: anishay2020@gmail.com

Twitter: @Physics_for_All